CLIMATE CHANGE AND AUSTRALIA

CLIMATE CHANGE AND AUSTRALIA

Warming to the Global Challenge

Ben Saul
Steven Sherwood
Jane McAdam
Tim Stephens
James Slezak

THE FEDERATION PRESS
2012

Published in Sydney by
The Federation Press
PO Box 45, Annandale, NSW, 2038.
71 John St, Leichhardt, NSW, 2040.
Ph (02) 9552 2200. Fax (02) 9552 1681
E-mail: info@federationpress.com.au
Website: http://www.federationpress.com.au

National Library of Australia Cataloguing-in-Publication entry
Climate change and Australia: warming to the global challenge /
Ben Saul ... [et al].

Includes index.
ISBN 978 186287 872 3 (pbk)

Climatic changes - Social aspects - Australia.
Global warming - Social aspects - Australia.
Emigration and immigration - Climatic factors.
Climatic changes - Government policy - Australia.

363.73874

Typeset by The Federation Press, Leichhardt, NSW.
Printed by Griffin Press, Adelaide, SA..

Contents

Acknowledgements

The authors thank their research assistants, Trina Ng (UNSW Law School), Naomi Hart (Sydney Law School), Laura Thomas (Sydney Law School) and student interns from the Sydney Centre for International Law at the Faculty of Law, The University of Sydney.

Table of Acronyms

ABARES	Australian Bureau of Agricultural and Resource Economics and Sciences
APP	Asia Pacific Partnership on Clean Development and Climate
AusAID	Australian Government Overseas Aid Program
BSF	Indian Border Security Force
CBD	Convention on Biological Diversity
CCS	carbon capture and storage
CDM	Clean Development Mechanism
CFCs	chlorofluorocarbons
CO_2	carbon dioxide
CO_2e	carbon dioxide equivalent
COP	Conference of the Parties
CPRS	Carbon Pollution Reduction Scheme
CSIRO	Commonwealth Scientific and Industrial Research Organisation
DFAT	Department for Foreign Affairs and Trade
ECOSOC	United Nations Economic and Social Council
ETS	emissions trading scheme
GATT	General Agreement on Tariffs and Trade
GDP	gross domestic product
HCFCs	hydrochlorofluorocarbons
IDPs	internally displaced persons
IPCC	Intergovernmental Panel on Climate Change
MOP	Meeting of the Parties
Mt	megatonne
NATO	North Atlantic Treaty Organization
NGO	non-governmental organisation
OHCHR	Office of the High Commissioner for Human Rights
ppmv	parts per million by volume
REDD	Reduced Emissions from Deforestation and Degradation
SO_2	sulphur dioxide
TPS	Temporary Protected Status
UNFCCC	United Nations Framework Convention on Climate Change
UNCLOS	United Nations Convention on the Law of the Sea
WTO	World Trade Organization
UNHCR	United Nations High Commissioner for Refugees
UNDP	United Nations Development Program
UNAIDS	Joint United Nations Programme on HIV/AIDS
WHO	World Health Organization

Table of International Law Instruments

	Climate Change Instruments
1992	United Nations Framework Convention on Climate Change (UNFCCC)
1997	Kyoto Protocol to the UNFCCC
2001–05	Marrakesh Accords (establishing the Clean Development Mechanism and other rules for implementing the Kyoto Protocol)
2007	Bali Action Plan
2008	Asia Pacific Forum, Niue Declaration on Climate Change
2009	United Nations General Assembly Resolution, "Climate Change and its Possible Security Implications"
2009	Copenhagen Accord
2010	Ambo Declaration
2010	Cancun Agreements
	Other Environmental Instruments
1972	Declaration of the United Nations Conference on the Human Environment (Stockholm Declaration)
1972	Convention on the Prevention of Marine Pollution by Dumping of Wastes and Other Matter (London Convention)
1985	Vienna Convention for the Protection of the Ozone Layer
1987	Montreal Protocol on Substances that Deplete the Ozone Layer
1992	United Nations Declaration on Environment and Development (Rio Declaration)
1996	Protocol to the Convention on the Prevention of Marine Pollution by Dumping of Wastes and Other Matter (London Protocol)
	Other Instruments
1947	General Agreement on Tariffs and Trade
1951	Convention relating to the Status of Refugees
1954	Convention relating to the Status of Stateless Persons
1959	Antarctic Treaty
1972	Convention Concerning the Protection of the World Cultural and Natural Heritage
1982	United Nations Convention on the Law of the Sea
1998	United Nations Guiding Principles on Internal Displacement
2000	Millennium Development Goals, found in the United Nations General Assembly "United Nations Millennium Declaration"
2001	European Union Temporary Protection Directive
2009	African Union Kampala Convention on Internally Displaced Persons

Introduction

Warming to a Global Challenge

Climate change is one of the most hotly debated issues of this century. Politicians have risen and fallen on its wake, elections have been fought around it, and international leaders have come under unprecedented scrutiny by civil society for their action – or inaction – in addressing it. As Australia is battered by hotter summers, more frequent and severe floods, devastating bushfires, water shortages in capital cities, and coastal erosion on its beaches, the question 'is climate change the cause?' is being asked more and more often.

The media is full of worrying projections. We are told that planned global emissions cuts will fall far short of the level necessary to cap temperature rises at 2°C (the so-called 'safe' level). We hear that the acidity of the world's oceans is increasing at the fastest rate in the last 55 million years, threatening disaster for marine life and food supplies around the world. Crop production and clean water may rapidly diminish and animal and plant species become extinct. Some reports boldly predict that up to a billion people will be homeless by 2050.

On the one hand, stories like these can encourage people to lobby for reductions in greenhouse gas emissions and do all they can to reduce their own carbon footprint. On the other hand, they can be so overwhelming that they create an almost paralysing fear, which can lead to a sense of fatalism and inaction. Climate change 'sceptics' create further confusion by suggesting that climate change is a hoax or conspiracy and that there is nothing to worry about. Occasional inaccuracies in some scientific studies have fuelled climate change 'denial', as if a handful of mistakes negates the thousands of studies which confirm the reality of climate change. Unfortunately, sceptics have succeeded in seriously misleading the public. According to a large public opinion survey of 5000 Australians in 2011, while 83 per cent of people agree that climate change is happening, only about 50 per cent believe it is human-caused.[1]

The gap between public opinion and scientific reality is stark. The Intergovernmental Panel on Climate Change (IPCC), the largest group of the world's leading scientists ever assembled – thousands of them – agree that climate change is happening and that human greenhouse gas emissions cause it. Getting that many of the world's most eminent and competitive scientists to agree is proof enough. Independent studies have also estimated that 97 per cent of scientific experts agree that humans are responsible for global warming,[2] corroborating the consensus in the peer-reviewed literature (that is, scholarly research papers which have been accepted by publication after being reviewed by experts).[3]

Any person who doubts that profound consensus would need to have a great deal of confidence in their own scientific ability, or be recklessly willing to gamble with the future of coming generations. Even if there is a remote chance that the scientific consensus is wrong, the likely serious impacts of climate change call for a 'precautionary' approach to contain the risks – rather than polluting as usual and hoping for the best.

The gap between public opinion and scientific reality is disturbing, particularly in a developed country such as Australia where there are high levels of literacy and education, a vibrant media, and easy access to information via the internet. It is difficult for governments to make good public policy to address a serious problem like climate change if a large section of voters has been misinformed – whether by climate sceptics, lobbyists, some academics, media propagandists, polluting companies, or political leaders who peddle misinformation for votes.

This book acknowledges where there are indeed uncertainties in the science, but pays no serious attention to 'sceptics' who deny that global warming is real, or that it is caused by humans. Given the near-universal scientific consensus to the contrary, those propositions are like believing that the world is flat, not round; or that gravity does not exist; or that the Sun orbits the Earth, not vice versa. Such denials of the basic fact of human-induced climate change are, in short, ridiculous, and not worth wasting more time on – even if some media outlets too often report them as if they might be true. At the same time, by focusing on mainstream science, this book equally discounts unfounded alarmist predictions of imminent global catastrophe.

It is a fact that the average global temperature has increased by 0.74°C over the past century. Historical greenhouse gas emissions will involve further warming of 0.6°C by 2100 even if emissions remain the same as in 2000.[4] If emissions continue to increase however, global temperatures could rise by up to 4°C by 2100 – with severe global effects that would change life as we know it. There is less certainty about *how fast the climate will change* (because it is difficult to predict the amount of future emissions and the scale of human efforts to mitigate them), *the impact of it on the planet's ecosystems and human societies* (taking into account natural resilience and human adaptation), and *what can be done to respond most effectively and efficiently*. This book is also concerned with exploring possible answers to these hard scientific and public policy questions.

This book was conceived of in the lead-up to the intense global climate change negotiations at Copenhagen in 2009. The world watched whether its leaders could put their own political, economic and cultural differences aside to act in the interests of the planet, and the well-being of future generations, by curbing global greenhouse gas emissions. International negotiations are, by their nature, forums for compromise, yet the achievements at Copenhagen were modest indeed – and a source of great disappointment for many. In Australia, too, attempts to introduce a national emissions trading scheme between 2007 and 2010 were unsuccessful, despite widespread popular support for action against climate change, and even helped to provoke the fall of Prime Minister Rudd. The Gillard government's announcement of a 'carbon price' in 2011 (see Chapter 4) met with fierce controversy and sorely tested the government's resolve before a new law was finally passed in late 2011.

Overview of this book

As political momentum for action stalls internationally and in many countries, this book aims to provide a clear, readable account of what climate change means for the future of Australia, its region and the world. Many books and articles about climate change take polarised positions, which can lead to uncertainty about what it is and what it means. This book takes a sober look at the science on climate change, explaining how it is already affecting our planet and continent, its future impacts, and the policy options available to Australia and the world in responding to it. Importantly, it gives a balanced and rational account of the diverse ways in which climate change impacts on society, drawing on the expertise of its authors in science, economics, geography and law. It is aimed at readers who are aware of and interested in climate change as a scientific, environmental, social, political and legal challenge, but it does not assume expert knowledge.

At the outset, Chapter 1 examines the science of climate change. It explains the processes of global warming, including the problem of human carbon or greenhouse gas emissions. It explores those aspects of climate science on which most scientists agree, and acknowledges areas where there is real (not fabricated) disagreement. It demonstrates that global warming is unequivocally real, serious and rapidly getting worse. It then explains the key impacts of climate change on the planet, including the atmosphere, oceans, polar regions, plants, animals and humans. It analyses the assessments of the IPCC, together with more recent science indicating that climate change impacts are happening even faster than predicted.[5]

Chapter 2 explores the physical and economic impacts of climate change specifically on Australia and its region. In particular, the chapter examines the special impacts of climate change on the diverse and vulnerable environments and ecology of Australia, as well as its likely effects on agriculture, industry, water resources and the livelihoods and prosperity of Australians. It also seeks to understand the possible economic consequences of such effects, and the extent to which they are offset by any positive impacts of climate change.

Chapter 3 explains how the international community has struggled to construct an adequate international response to climate change. It explains the key international treaties, such as the UN Framework Convention on Climate Change (UNFCCC) and the Kyoto Protocol. It evaluates their success, and analyses what is needed in a post-Kyoto world. In particular, it examines the outcome of the Copenhagen (2009), Cancún (2010) and Durban

(2011) negotiations, and charts possible future directions for global cooperation. It explains why countries need to adopt a coordinated global response to climate change. It also looks at other areas of international law that are relevant to addressing climate change, including environmental law.

Chapter 4 moves from the global to the local by exploring Australia's response to climate change. It examines Australia's mixed record in international climate negotiations, before turning to look at how well successive Australian governments have responded in their domestic policy choices. It shows that Australia has many options open to it for reducing its own carbon emissions, yet has been relatively slow in taking them up. The reasons why Australia should reduce its emissions are also considered, in the light of global efforts, the nature of the Australian economy, and Australia's responsibility for a high level of per capita emissions and a relatively 'dirty' economy. Central to the policy debates was the impasse over the introduction of an emissions trading scheme between 2007 and 2011. This chapter looks at the costs and benefits of 'emissions trading' (compared with other options, such as a carbon tax, or direct action), and explores the hard issues about its design. It also shows how many of the criticisms of carbon pricing have been misleading or exaggerated.

The final two chapters of the book examine particularly complex challenges presented by climate change. Chapter 5 confronts alarmist predictions that up to 250 million people – many in the Asia-Pacific region – may be displaced by the effects of climate change by the year 2050. The chapter queries the extent to which empirical evidence supports such claims, and whether climate change can be isolated as a cause of such movement. It examines the extent to which existing international refugee and human rights law would protect so-called 'climate refugees', and where the legal gaps exist. In particular, it considers the ways that planned migration may help to avert crises, and assist people to re-establish their lives with dignity. It looks at the role Australia might play in assisting people in its region, such as Pacific islanders, whose homes may be rendered uninhabitable as a result of drought, water shortages, king tides and eventual sea-level rise.

Chapter 6 finally considers predictions that climate change will lead to conflict, violence and even 'climate wars'. It argues that, although no war is likely to occur on account of climate change alone, its cumulative effects may aggravate existing causes of violence, worsening existing political, social or economic vulnerabilities and undermining human security. Already, declines in global food production, partly because of climate change harming agriculture, have triggered food riots and social conflict in developing countries and destabilised governments. Water scarcity, public health problems, human displacement, and resource competition over land and maritime and polar areas, will all present special challenges. This chapter examines the extent of these risks and considers how the international community – including Australia – can best respond to avert conflict or tensions stemming from the impacts of climate change.

The global economy of the future will look nothing like the carbon-intensive economy of today, on which Australia so heavily relies. There are certainly adjustment costs in moving from a 'dirty', carbon-polluting economy to a 'cleaner', more modern one, and reasonable concerns about Australia sharing a fair burden of reducing global emissions. Some clearly have vested interests in maintaining the status quo and preserving their freedom to pollute – a bad habit from an era where those who caused environmental damage were never asked to pay for it. Political leaders face acute difficulties in planning for such a long-term problem as climate change, when electoral cycles are so short, voters are so cynical, and opposition parties are so rapacious.

But there are great rewards available to countries that are bold enough to make the transition. Countries which lag behind will be less able to reap the benefits of new employment, exports and investment in new technologies and industries. Even if Australia did more than other countries to reduce its emissions, that would not necessarily be a bad thing – apart from showing moral courage and leadership, it could also aggressively position Australia's economy for future success. Dealing with climate change is not just a narrow environmental problem, but challenges the way we think about our economy, security and way of life. Responding to it should not be feared, but embraced as an exciting and productive opportunity for renewed global cooperation, and good Australian policy to secure our future.

Notes

1 CSIRO, *Baseline Survey of Australian Attitudes to Climate Change: Preliminary Report*, January 2011.

2 Doran, PT and Kendall Zimmerman, M, 2009, 'Examining the Scientific Consensus on Climate Change' 90 *Eos, Transactions, American Geophysical Union* 22; Anderegg, WRL, Prall, JW, Harold, J and Schneider, SH, 2010, 'Expert Credibility in Climate Change' 107 *Proceedings of the National Academy of Sciences* 12107.

3 Oreskes, N, 2004, 'Beyond the Ivory Tower: The Scientific Consensus on Climate Change' 306 *Science* 1686.

4 UN Secretary General, 'Climate Change and Its Possible Security Implications', Report to the UN General Assembly, 11 September 2009, UN Doc A/64/350, para 23.

5 See, eg, Steffen, W, *Climate Change 2009: Faster Change and More Serious Risks*, Australian Government, Department of Climate Change, Canberra, 2009.

Chapter 1

A Crisis in Slow Motion: The Science of Global Warming and the Response Needed

Key Issues

- The Earth's climate has undergone numerous natural changes over millions of years. The geologically recent climate has been unusually stable, allowing humanity to flourish.
- Over the past 250 years, the global average temperature has risen by nearly 1°C, with the fastest increases reported during the past 30–40 years. The adverse effects of this increase are most evident in sensitive parts of the environment, including forests, polar ice sheets and glaciers, and for human populations living in extreme climatic conditions.
- Increases of carbon dioxide, methane and chlorofluorocarbons in the atmosphere trap heat, causing a rise in global temperatures. Excess carbon dioxide does not break down easily in the environment and takes hundreds of thousands of years to be absorbed.
- Not all variations in the Earth's climate or temperature can be attributed to greenhouse gas releases, and not all greenhouse gas emissions are human-made. Over the Earth's history, changes in solar, volcanic and geological activity have caused climatic variations.
- There is overwhelming scientific agreement that human carbon emissions are great enough to disturb the natural cycles of carbon that have previously maintained a relatively stable global climate.
- By 2100, temperatures are predicted to increase by 2–6°C, reflecting a doubling of atmospheric carbon dioxide levels. These predictions are supported by actual rises in temperature over the 20th century. Even an increase of 2°C would dwarf any other temperature change since the stone age.
- Once warming occurs, it almost certainly provokes oceans and land-based ecosystems to release some of their own carbon, exacerbating the problem.

- Both the maximum eventual warming and ocean acidification will be determined by the total amount of carbon dioxide *ever* emitted, irrespective of when or at what rate those emissions occur. This makes curbing emissions an urgent global priority. Humans have already emitted about half the total amount that is likely to produce a 2°C rise.
- It is difficult to predict the effects of global warming. Sea-level will rise, though how quickly is uncertain. Global average sea-level already rose 20 cm between 1870 and 2007. Low-lying island nations may be wholly submerged, and the populations of coastal regions may have to relocate inland. Global warming will make oceans more acidic, and shift rainfall patterns and climate zones, in turn damaging fragile ecosystems, lessening biodiversity and impeding food production, especially in tropical and polar regions.
- Societies will adapt to an altered environment, but this is more difficult if changes are larger or more sudden, or if populations are already stressed. In the short-term, the most effective ways to curb emissions growth are through energy conservation and the adoption of new energy technologies. Government policies also need to encourage or compel changes in behaviour, lifestyle and technology use.
- The most serious and urgent changes are required in energy consumption patterns, especially in lighting, heating and cooling, and transport. Using natural gas, nuclear power and renewable energy, rather than coal, will help slow global warming.
- If emissions are not reduced, more radical 'geo-engineering' strategies (such as artificially removing carbon dioxide from the atmosphere or creating means to reflect sunlight away from the atmosphere) may be necessary, but their effects may be unpredictable.
- Many economists have considered the costs of mitigating climate change against the costs of doing nothing. Some of the costs of climate change are likely to be unforeseeable. The *Stern Review* in the United Kingdom estimated that stabilising greenhouse gas levels at acceptable levels of carbon dioxide would cost around 1 per cent of annual global gross domestic product (GDP) by 2050, which is probably small compared to the costs of inaction.

A Introduction: The discovery of a problem

In 1938, the engineer and amateur scientist Guy Callendar stood before the Royal Academy in London and proclaimed that increasing carbon dioxide (CO_2) in the Earth's atmosphere was causing the world's climate to warm. He pointed to calculations done – in 1896 – of the impact of this gas on the planetary heat balance, and his own analysis of past measurements of CO_2 and temperature around the world. His data was sketchy but seemed to agree with the calculations. Colleagues were mostly sceptical.[1] Seventy-odd years and another 0.5°C later, the reality of this phenomenon is still sinking in, despite the vast amount of additional evidence to uphold his claims.

Earth's climate has been through a lot of changes. The further back in time one goes, the more uncertain the information, but there is geological evidence of periods early in the Earth's history when the surface was nearly ice-covered. During and after the time of the dinosaurs, the planet was hot enough that crocodiles could live in the Arctic circle, palm trees grew in Wyoming and Antarctica was covered by a thick forest. About two million years ago, after cooling down again, the planet began gyrating in and out of ice ages that have caused sea-level changes of over 100 metres and great expansions of polar ice.[2] Earth's climate is obviously subject to change. It was curiosity about what drove the ice ages that provoked Callendar and many others to first investigate the climate.

The climate in modern times has been unusually stable. The present period, called the Holocene Epoch by geologists, began with the end of the last ice age about 12,000 years ago and there has been little movement in key indicators since then. Greenhouse gas concentrations, analysed through ice core records, had barely changed by the time Captain Cook landed in Australia in 1776. The average global temperature at the beginning was probably within 1°C of what it was at the end of World War Two. Sea-levels rose by only a few metres over that period, and not at all for the past 2000 years.[3]

These placid conditions have enabled humanity to flourish. At the end of the last 'ice age', the human population was probably about five million nomadic hunter-gatherers whose most sophisticated tool was a stone axe. Today there are seven billion people on Earth and this number is rising by about 80 million each year.[4]

In the past, different civilisations have taken a few knocks from regional climate changes. For example, the Viking colony in Greenland starved during the 'little ice age' around the 17th and 18th centuries, when a chill over the region put a brutal end to their experiment in Arctic agriculture. Prolonged regional droughts are also recorded, apparently contributing to the collapse of several civilisations including the Mayan in Central America. While many assume that such changes were global events, and sceptics hold them up as examples of natural global warming, careful study of tree rings and other records shows a more complex picture.[5] When the climate changed in one place, other places experienced changes that if anything were small or opposite.

Since the beginning of the industrial revolution in Europe about 250 years ago, however, a new kind of change has begun: the coordinated increase of temperatures nearly everywhere on Earth. The global average temperature has now risen by nearly 1°C, with the fastest increases reported during the past three to four decades. Increases have occurred on every continent and in most of the world's oceans hundreds of metres below the surface.[6]

Table 1.1: Some different climates throughout history

Period	Global temperature relative to 1850	Approximate CO_2, ppmv (280 in 1850)
Cretaceous (c. 100 Ma)	+10-15 C	600-2400
Eocene (c. 40 Ma)	+5-12C	300-1500
Oligocene (c. 10 Ma)	+3-5C	250-1000
'Ice age' (c. 20 Ka)	–5-6 C	180
Medieval warm (c. 1 Ka)	+0-0.5 C	280
Little ice age (c. 300 A)	–0-0.5 C	280
2010	+0.8 C	390
2100 (business as usual)	+3-7 C	1000

Note: Ma = million years ago; Ka = thousand years ago; A = years ago

One degree of warming is not something the average person would directly notice. But if sustained, it can be seen in parts of the environment that are more sensitive: forests, polar ice sheets and glaciers, permafrost and sea-levels. All of these are now showing unmistakable warming signs.[7] Also, some people are more attuned to their environment or have experienced greater changes. To the indigenous Inuit people living in northwest Alaska, for example, climate change is obvious to anyone over the age of 30. Frostbite has become a thing of the past; weather is becoming more variable; sea ice is shrinking; midwinter fogs are appearing for the first time in memory; and herds of reindeer that have been hunted for many generations have moved northward, replaced by caribou and moose. Similar stories can be heard throughout the greater Arctic region.

In Australia, some critics claim that winters (or summers) were colder (or hotter) when they were growing up. Generally these imperfect memories have nothing to do with real climate change. One change, though, is obvious: the increasing dryness in many southern regions. This appears is partly due to human activity, but proving that is not always easy.

B The great climate science debates

The claims of Callendar and others sparked studies and eventually heated debates on all aspects of the climate problem. Carbon dioxide, water vapour and methane had already been identified as leading greenhouse gases - gases that absorb the infrared radiation which the Earth must emit into outer-space in order to balance the energy it gains from the sun. Scientists realised in the 1800s that the surface would be considerably colder (some 33°C according to modern analyses) without these gases,[8] but it was not clear how much difference more of them would make. In the 1960s, physicists developed solid spectroscopic theories, backed by laboratory measurements, with which they could finally calculate an answer: additional gas would have less impact on climate than the gas already present, but still enough to matter given the huge amounts being generated.

These calculations have held up over time. In fact, there are now satellites that can directly measure the greenhouse effect of each gas by observing its fingerprint in the outgoing spectrum. These satellites confirm that the effects of CO_2 and methane are growing as expected and confirmed the growth of a new greenhouse gas: chlorofluorocarbons (CFCs), the refrigerant chemicals more famous for destroying the ozone layer.[9] Carbon dioxide is trapping 200,000 Gigawatts more power today than it was 30 years ago, with smaller increases for the other gases. There is no longer any doubt: rising greenhouse gases are heating the planet.

The next question was whether human emissions would be enough to perturb the vast natural cycles of carbon. Until the end of the 1950s, most believed that human carbon emissions would be quickly absorbed by the oceans and land surface, which contain far more carbon than the atmosphere. Unfortunately, oceanographers discovered two problems. The ocean turns over so slowly that it would take at least a millennium for most of the available water to be brought into contact with the air. Even worse, chemical buffering mechanisms strongly inhibit the uptake of carbon by seawater.

The implication – that human emissions would not conveniently disappear into the oceans – was soon confirmed when precise observations of carbon dioxide (that began in 1957) showed that about half the emitted carbon was accumulating in the atmosphere.[10] Very recent observations suggest that this fraction is growing as the oceans and land cannot quite keep up.[11] In any case, CO_2 is increasing at about 1 per cent per year.

At this point, two things are certain. Greenhouse gases cause warming, and human activities are increasing greenhouse gases. Many other types of evidence prove that this rising carbon comes from the burning of fossil fuels: its unique isotopic fingerprint, sudden appearance in the geologic record, and difference in concentrations between the two hemispheres.

It is commonly objected that climate has changed before, and nobody was driving SUVs then. So how is it known that this is not just another natural change? And if those changes were caused by greenhouse gases, from where did the gases come? Many past climate changes remain unexplained and more are being discovered;

this is a fascinating area of ongoing scientific research. However, calculations have shown that the largest swings in climate over the past half-billion years can be accounted for fairly well by CO_2 changes, sometimes combined with other influences.[12]

In particular, CO_2 amounts have drifted up and down due to gradual changes in volcanic and other geological activity. Sometimes they rise more rapidly due to releases from the oceans. There were also a few sudden jolts to the climate system and global carbon cycle, the most prominent causing at least 5°C of sudden global warming around 53 million years ago. Researchers are not sure what caused these spikes, but each of the suggested culprits – volcanic eruptions, massive peat fires or catastrophic methane releases – involves a greenhouse gas release.[13]

The world is not always simple, and not all major lurches in climate were greenhouse-caused, including the relatively recent glacial cycles. In fact, the ice core evidence suggests that carbon dioxide variations occurred slightly *after* the climate variations – as scientists had anticipated in advance – rather than before.[14] These cycles were clearly driven by changes in the Earth's orbit around the sun, which subtly redistributed the solar heating of the planet and caused global warmings and coolings. The details remain unclear, but each warming evidently caused carbon dioxide to emerge from the world's oceans and methane from land areas. This made the climate swings much larger than they would have been. It also explains how the redistribution of summertime sunshine to the *Arctic* caused the *southern* hemisphere to warm along with the northern one.[15]

What about the influence of the sun? Though some climate changes hint at having been caused by solar variations, this has been tough to confirm. Recent calculations have effectively explained the earlier 'medieval warm period' and 'little ice age' on the basis of the combined effects of changes in solar and volcanic activity.[16] While sceptics still occasionally blame the sun for modern warming, the evidence indicates only a feeble contribution, perhaps 5 to 10 per cent.[17] A greater influence on modern climate has been particulate air pollution, which humans began to generate in earnest when they discovered the convenience of coal for cooking and heating. This pollution tends to cool the planet. Without it, there would have been even more global warming before clean air regulations came in during the 1950s and 1970s.[18]

Politicians and sceptics continue to speak of these interesting but old debates as if they were still going on, because it allows them to avoid confronting the climate problem. The scientific community, which is lucky enough not to be responsible for solving the problem, has long since moved on to the real question: what are the consequences?

C Future warming: how much and how sure are we?

It is scientifically known that human activities are causing a rapid greenhouse gas buildup; that greenhouse gas increases raise the global temperature; that global warming has matched IPCC predictions so far, and warming will significantly worsen if it is business as usual. There is, however, uncertainty about the exact severity of likely future warming; how patterns of rainfall and other changes will manifest; and what will be the impacts of warming.

How much future warming is likely to occur? There are several ways to answer this question. Past climate records can be analysed to estimate how much average temperature increases under a given rate of sustained additional heat input (say, 200,000 Gigawatts; to put that figure in context, a large Australian coal-fired power station is around two Gigawatts). This important number is called the 'climate sensitivity'. An alternative way to estimate it is from theoretical models of climate, but a big problem is that the answer depends on subtle changes that might happen to, for example, clouds. That is a hard problem. One can simply assume that clouds do not change, or else simulate future climate with a 'general circulation model', an elaborate computer program that does its best to predict all aspects of global weather including the clouds.[19] General circulation models are also the only way to predict important regional phenomena – say, future rainfall in the Murray-Darling basin.

While none of these three approaches provides an exact answer, all of them point to the same range: somewhere from 2–5°C warming for the heat input caused by a doubling of CO_2. Given expected global emissions growth, this leads to about 2–6°C more warming by 2100, with even more after that. Again, sceptics point to outlier studies suggesting a much less sensitive climate (ignoring other outlier studies at the opposite extreme). But they cannot then explain the huge changes in the geological climate record,[20] and their arguments appear to be based more on wishful thinking than on reason.

One factor supporting the model predictions is that, when the same models are used to simulate the 20th century, the results look pretty close to what actually happened. This goes not only for the average warming rate, but the pattern of more warming in winter and in polar regions and cooling in the stratosphere; and other changes, like increasing humidity.[21] Some of these aspects are unlike what would have been expected if the changes were driven by other factors like the sun. Moreover, solar output has not budged since good measurements began in 1979, yet warming has been as rapid as ever.[22] It is fairly clear that most 20th century warming was human-made.

One could object that the modellers already knew the answer when they built their models – but the predictions published by the IPCC since 1990 have very closely tracked the warming since then.[23] To use a medical metaphor, a disease was first diagnosed and its symptoms predicted. The symptoms then appeared as predicted, and new tests confirmed the original diagnosis. Making the case even stronger, further studies have identified no alternative conditions that could cause exactly the same symptoms.

The proportion of experts that agrees that humans are largely responsible for observed global warming has been measured at 97 per cent by two independent methods,[24] corroborating the consensus in the peer-reviewed literature.[25] Many esteemed scientific bodies have issued statements on the science, notably the Australian Academy of Science.[26] It is not impossible that this consensus could someday be fragmented by new evidence, or that the predicted amount of future warming could be revised. But if so, the predicted change could just as easily become worse as better. In fact this, if anything, has been the trend so far as knowledge has progressed.

Other issues arise in predicting the future. Once warming occurs, it almost certainly provokes oceans and land-based ecosystems to give up some of their own carbon, adding to the problem. But how much? Again, past variations may be a more useful guide than theory: recent evaluation of climate records over the past two millennia suggest about 20 parts per million by volume (ppmv, a measure of the concentration of a gas) of extra CO_2 are released due to each degree Celsius of warming, or enough to amplify warming by up to 25 per cent.[27]

All bets are off, however, if warming goes outside the envelope of past changes, as it is likely to do later this century. This might push forests or other ecosystems too far, causing massive carbon releases – one example of a likely 'tipping point' in the system, and an important reason to moderate future warming even if it cannot be prevented outright. This is just one example of the uncertainty in gauging the seriousness of global warming.

D The real problem, and why 'wait and see' won't work

While several factors have influenced climate so far, looking into the future one stands out: CO_2. The main reason is that it does not break down in the environment. Instead it must be absorbed into sediments by a painfully slow process that takes over 100,000 years to reach equilibrium.[28] By contrast, methane breaks down chemically in about a decade, and even CFCs last only a century or so. Water vapour is removed in a matter of days, making emissions of this gas irrelevant, and air pollution is removed in a week or two. But once fossil carbon is released into the climate system, much of it is essentially there forever as far as humans are concerned (unless it is somehow artificially removed). Other influences on climate will eventually pale in comparison.

The urgency of the CO_2 problem follows from some simple facts. Both the maximum eventual warming and ocean acidification are determined by the *total amount of* CO_2 *ever emitted*, almost independently of when or how fast.[29] This means our intuition from familiar pollution problems does not apply. With most pollutants, when polluting is stopped, the problem goes away before long. This means it may be reasonable to wait until a problem is clear and significant before acting. But with CO_2, stopping the emissions does not make the problem go away, it only stops the problem from getting any worse.

Stabilising emissions is like taking your foot off the accelerator if your car is heading for a cliff. It is a step in the right direction but, unless the cliff is very far off, it is not enough - you need the brakes. In addition, because CO_2 emissions are so pervasively woven into the modern economy, changes will take decades to implement. Imagine you are heading toward a cliff and the brake pedal is hidden somewhere in the car.

Unfortunately, humanity still has its foot on the accelerator. Humans have already emitted about half the total amount that is likely to produce a 2°C temperature rise, and most of it since 1970.[30] The amounts of CO_2 emitted are staggering: a person who owns a typical car and drives 15 kilometres to work each way puts about eight kilograms of CO_2 into the atmosphere with each return trip. Human CO_2 emissions to date have probably already been enough to prevent the next ice age, which would otherwise have come on gradually over the next 80,000 years.[31] Chart 1.1 shows the sources of human carbon emissions globally.

The coming warming is likely to be big, but not catastrophic if humans can find the brake pedal. With continued population and economic growth fuelled mainly by the exploitation of new fossil fuels, CO_2 levels would skyrocket to roughly 1000 ppmv by 2100.[32] This is more than triple the pre-industrial value of 280 ppmv, implying a global warming of 3–7°C by 2100 compared to the pre-industrial climate, with further warming in the 22nd century. The range of 3–7°C is wide to be sure, but even the low end represents a climate change dwarfing any since the stone age. If all available fossil fuels are burned up, the resulting warming could exceed 10°C.[33] With very strong mitigation measures, however, the warming by 2100 could be cut in half and, more importantly, further rises could be stopped.[34] This could spell the difference between catastrophe and inconvenience for people living a century from now.

Chart 1.1: Global carbon emissions by sector, 2007 (%)

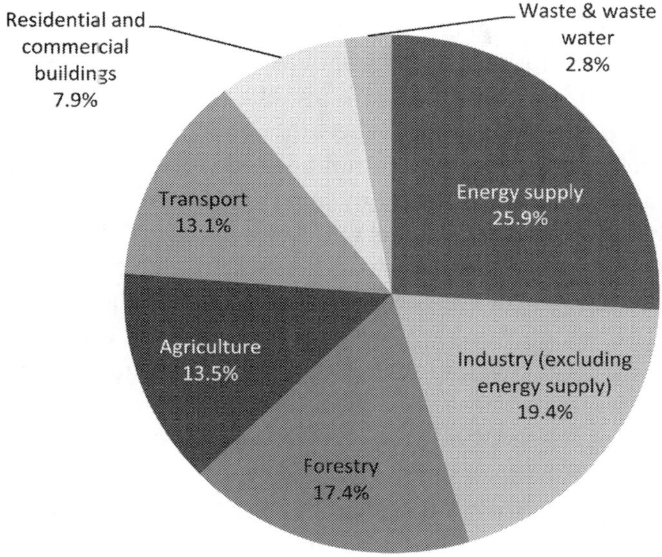

Source: Intergovernmental Panel on Climate Change, Contribution of Working Group III to the Fourth Assessment Report of the Intergovernmental Panel on Climate Change: Technical Summary, 2007, p 29.

E What would a warmer future be like?

Predicting global mean temperature is child's play compared with predicting what warming will do to societies or ecosystems. Past examples of when the earth warmed or cooled naturally are relevant, but these are hard to turn into concrete predictions of the future because the imminent global warming rate, the current human population and the fragmentation of natural ecosystems are each unprecedented.

As noted in Chapter 3, the international community recently agreed to recognise 2°C of global mean warming (above the temperature in 1850) as a threshold for 'dangerous' climate change. There is no obvious scientific justification for this and it is a political number as much as a scientific one. One reasonable rationale, however, is that +2°C corresponds roughly to the upper limit of temperature changes recorded in ice cores during the past million years. So there can be confidence, for example, that neither the Greenland nor Antarctic ice sheets would melt with 2°C warming even if sustained for millennia. Warmth in excess of that, however, takes us into territory not seen for millions of years when the great ice sheets of today were not always present.[35]

Sea-level rise is one impact where the geologic record indicates roughly *what* to expect, though not *when*. Sea-levels throughout the Earth's past have fluctuated by hundreds of metres, and were generally higher when the planet was warmer. This is because of at least two things: sea water expands when heated, and land ice melts into the oceans. At the end of the last ice age, when the earth was only 5°C colder, the sea-level rose 120 metres mainly due to melting of an ice sheet covering what is today Canada and part of the United States. Antarctica still holds 60 metres worth of additional sea water, but this too was in the oceans once, when the planet was about 10°C warmer than today.[36]

Thus, eventual rises of at least five metres per degree seem guaranteed. That is 15–30 metres for the 3–7°C warming due in 2100 with business as usual – enough to inundate most of Sydney. But it would almost certainly take centuries for the ice to melt, somewhat like a big ice sculpture on a hot summer's day. The rise expected by 2100 is about one metre.[37] Global average sea-level already rose 20 cm between 1870 and 2007, at a rate of 1.7 mm per year during the 20th century and an accelerated 3.7 mm per year between 1993 and 2007.[38]

The rising seas may gradually cause low-lying island nations to disappear. The total number of displaced people will be modest (for example, the current population of Tuvalu is only about 10,000 people) and other factors like land-use are partly responsible for the problems these islands are now experiencing (see Chapter 5). Nonetheless, there is symbolic, cultural and philosophical significance in losing even one small country. While most countries will not lose substantial portions of their land area to sea-level rise, much of the built and natural infrastructure near the coast will be threatened and many people live near the current sea-level. In many tropical areas, mangroves play a crucial role in sheltering coastal regions from the impacts of cyclones, but sea-level rise will gradually destroy established mangroves.[39] Adaptations to establish new mangroves further inland, and to move the built environment inland, will be necessary to limit the damage.

The oceans are changing in another way: they are growing more acidic, due to CO_2 uptake. Soaking a dirty coin (or a seashell) in soda water or a soft drink demonstrates what CO_2 can do in liquid water. This completely separate impact of emissions is unrelated to climate but potentially as serious. While CO_2 levels in the geologic past were much higher than today, they changed slowly enough for chemical buffering mechanisms to limit the impact on shell-building organisms. But the current input of CO_2 is much more explosive. The *p*H level of the upper ocean has already fallen on average by 0.1 units in the past century, a rate probably unequalled in over 55 million years.[40]

Where this will lead is uncertain but it is already affecting some marine life. Coral growth is slowing; these and shell-building organisms will eventually find conditions intolerable if *p*H falls low enough, though ecosystem tolerances and adaptabilities are poorly understood.[41] Impacts on reef, planktonic and other micro-organisms can propagate up the food chain to affect species of direct value to humans, including fisheries.

On land, shifting rainfall patterns and climate zones present one of the most worrisome threats from climate change. Because rainfall patterns are so variable, and because available indicators of past rainfall like tree rings are far from perfect, it is hard to say much about the future from looking at these records. But there are good theoretical reasons to expect rainfall in a warmer climate to become more intense but less frequent.

This trend is also predicted by the general circulation models, and was beginning to appear in many rainfall records worldwide even before the disastrous floods in Australia and elsewhere around the world in January 2011.[42] This is usually bad news since rain is generally more useful and less damaging when it is steadier rather than highly episodic. More importantly, climate models and basic reasoning predict that the subtropical belts on Earth will become even drier than they are now, while high latitudes and the deep tropics will receive more rain.[43] This is significant for Australia (see Chapter 2).

All of these changes will surely affect the health of ecosystems and biodiversity around the world. The rate of change is a key factor: if change happened slowly enough, the plant and animal communities could gradually migrate while maintaining biodiversity. The ecosystems at the end might be different, but equally valuable. But a warming of several degrees in only a century or two would leave most ecosystems stranded and unable to cope with alien conditions.[44] Imagine an oak forest trying to move several kilometres per year for many decades. Even that assumes that the same conditions occur, just farther poleward. In reality, the delicate balance of rain and temperature cycles needed by a particular ecosystem might vanish altogether.

If the climate changes too much, forests or other plant communities will die off – possibly via bushfires – releasing stored carbon back into the system. This is something to think about when considering tree planting as an emissions abatement strategy. The likelihood of die-offs increases significantly above 2°C.[45] The new (often weedy) systems that become established may be significantly lacking in biodiversity and capacity to produce 'ecosystem services' for humankind. It could take millennia for new ecosystems to redevelop their natural richness. Some hybrid climate-biosphere models predict that 3–4°C of warming would kill the Amazon rainforest and replace it with savannah, although most models predict that this will not occur, or would require 6°C of warming.[46]

F Who will be most affected by climate change?

Heat impacts on ecosystems and humans, while hard to predict, are certainly going to be more of a problem in the tropical and polar regions than in-between. Not only are the tropics already warm, but temperatures vary so little there that even modest warming would be significant.[47] Corals are a well-known example of this – a wondrous organism but a true 'hothouse flower' that can tolerate little warming. While flora and fauna at mid-latitudes can migrate poleward or to higher elevations – and hundreds of species are indeed doing so now – this does not help polar bears or tropical plants, for example. Polar creatures have nowhere colder to go. Tropical ones could move poleward, except they would never survive in the vast subtropical dry zones that would block their way.

With a warming of 2°C or more, tropical heat waves will begin to surpass anything experienced today, which will be tough on many of the plants and animals. This problem becomes significantly worse with 4°C of warming, at which point many of today's crops (such as wheat) and animal husbandry would likely not work out except at elevation.[48] Significant adaptations will be needed even with 2°C of warming, and it is not clear what would be possible with 4°C, where new heat-resistant crops would be needed. With 6°C of warming most of the tropics would likely be uninhabitable by humans due to heat stress, no matter what crops they had developed.[49]

Warming will bring profound disruptions at high latitudes, but also benefits in the form of longer growing seasons, milder winters and expansion of forests into areas that are now tundra. Thus some countries such as the United States, Canada and Russia may experience climate change as a mixed bag, at least for the first couple of degrees.[50]

Not only will developing countries at low latitudes be hit hardest by climate change, but their adaptive capacity is also much more limited. In the developed world it is not hard to imagine new crops or other technologies saving the day for farmers. But given that most farmers in the developing world do not even use tractors or fertilisers today, new technologies may not easily reach them. A lack of education or skills is as important a barrier as finance, and new stresses like climate change are a powerful reason to improve education in developing nations, not to mention developed ones.

If really large warming happens, the outlook is not good for countries in any economic category. Four degrees Celsius would have severe negative consequences for all, and greater warming (6°C) would leave few places on Earth pleasant to live in. Moreover, these greater warmings might drive an exodus of millions of people from tropical regions (see Chapter 4). So, despite the unequal impacts of the climate changes that will occur over the next few decades, climate change should not be seen as a problem only for the developing world. If untreated it will threaten all of humanity.

G What can be done about climate change?

Societies will adapt to climate change, but this becomes more difficult in the face of the larger warmings that are likely if emissions continue unabated. This indicates the need for 'mitigation' or interventions to reduce greenhouse gas emissions, and even 'geoengineering' or physical interventions to try to undo climate change.

In the short term, the effective ways to curb emissions growth are through energy conservation and adoption of new technologies. Efficient strategies will involve attacking not only emissions of CO_2 but also of methane, soot, ozone, CFCs and nitrous oxide, each of which warms the planet. In the long-term, however, CO_2 emissions would have to sink to nearly zero regardless of how much progress is made in reducing the others. This is a daunting prospect, but on the other hand fossil fuels are a finite resource anyway, so it is really a matter of moving the deadline forward.

The question of how to achieve the needed reductions breaks down into two. First, what changes in behaviour or new technologies are available that can be part of the solution? Secondly, what government policies or actions by other actors would encourage or compel people to do this? While the emphasis is often on technology, incentives and education are equally important.

A simple example shows why new technology alone is not enough. For years, car manufacturers have known many ways to make cars much more fuel-efficient, for example by making them lighter. The problem is that buyers do not want light cars, mainly because in a collision the heavier car wins. The result is effectively an arms race where everyone buys the heaviest car he or she can afford. Yet if everyone bought a lighter car, any car they collided with would also be lighter. Thus a simple policy decision to favour or require lightweight cars would save a lot of fuel without really sacrificing average road safety. The status quo is a classic market failure, where rational individual choices lead collectively to bad results.

Nearly all economists would agree that policies of one sort or another that penalise or prohibit carbon-intensive choices are essential for really driving down emissions.[51] The catch, of course, is that there is more than one way to make a car lighter; policy must be well designed so as to avoid, say, encouraging death-trap cars or those that are lighter but still do not save petrol. Anyone who followed the fiasco with the federal home insulation scheme in Australia knows the potential for well-intentioned but badly designed policies to misfire.[52] However tricky it may be to design them, they remain essential. Chapter 3 looks at how the international community has attempted to design a global agreement to coordinate action to reduce emissions. Chapter 4 looks more closely at the policy options available in Australia, and how they link into those international efforts.

Education is also important, since people do not always respond rationally to financial incentives. Many businesses have found that they can save money by cutting their carbon emissions. So why did they not do it before? Because they did not see the waste, not because there was no incentive to eliminate it. Consumers are even worse: they routinely turn down small up-front investments, such as home insulation, that would save big money down the track. Simply informing people of how much more energy-hungry they are than their neighbours, for example on their heating bills, can nudge them into conserving. Competition and non-financial recognition of environmental efforts can also have a big effect. Stores in the United Kingdom are beginning to label some products with their carbon footprints; ALDI supermarkets, in cooperation with the Planet Ark group, recently announced it would do so in Australia.

Assuming perfect information and full motivation, what then should be done, given currently available technology? The heart of the problem lies in energy use, which accounts for over three quarters of global CO_2 emissions. In most developed countries, energy use is led by electricity generation (mostly from natural gas and coal), followed by transportation (almost all from oil-derived fuels) and industry.[53] About 70 per cent of global electricity production comes from coal, gas and oil, while only 16 per cent comes from renewables and 14 per cent from nuclear power.[54] The energy sector also accounts for some methane emissions due to natural gas leaks, and soot from low-temperature burning sources such as cookstoves in developing countries. Chart 1.2 (*over page*) shows the key sources of global energy.

Chart 1.2 Global energy supply by fuel type, 2006 (%)

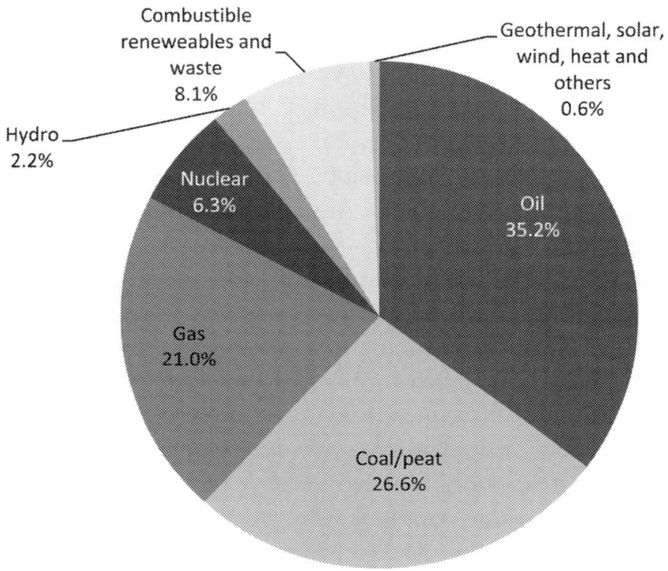

Source: International Energy Agency, Key World Energy Statistics, 2008, p 6.

The second-leading source of greenhouse gases is land use, primarily deforestation and agriculture (with much of the former done to provide land for the latter). This contributes approximately 20 per cent of CO_2 emissions, mainly through burning or decay of biomass which releases its stored carbon. Animal ruminants are a significant source of methane and fertiliser use generates nitrous oxide. A final source worth mentioning is cement production and natural gas flaring, which accounts globally for 4 per cent of carbon dioxide emissions (50 per cent more than aviation).[55]

So the lion's share of reductions will have to come from energy use. The options here include improved efficiency, conservation, and use of non-fossil sources or fossil sources with carbon capture and storage technology. Most or all of these options must be pursued simultaneously to achieve what is needed.[56]

On the efficiency side, there are too many options to list, as nearly every product could be more efficient. This text is being written on a laptop that one of the authors prefers in part because it uses much less electricity than standard desktop computers. But the big areas are heating and cooling of buildings, transportation and lighting. Improving building design and insulation are areas where the greatest value for money is possible. Australians are slowly discovering that insulation works for keeping heat out as well as it works for keeping it in. Strategies for reducing transportation emissions include redesign of cities or ways of reducing commuting distances; encouraging bicycles and walking, better public transport and, of course, electric or hydrogen vehicles. The catch with these is that, as long as electricity comes from coal, they are no better than standard cars.

Coal is now the leading source of electric power worldwide, and is also the dirtiest fuel available, generating more than double the CO_2 per watt of power as natural gas. So it would help significantly to replace coal with gas, at least until better alternatives can be developed. Unconventional gas reserves in frozen deposits and undersea clathrate reserves are likely to far exceed conventional ones.[57] Nuclear (fission) power is nearly carbon-free and can scale up to be a dominant electricity source, but has obvious security and environmental issues (as the Japanese earthquake of 2011 showed), a less-appreciated one being its thirst for water. It is also fairly expensive and is non-renewable, although Australia has amongst the world's largest uranium deposits.

Among renewables, wind energy is catching on fast and could generate 20 per cent of electricity with old-fashioned grids, or much more with investment in modernised 'smart' electricity grids. It is already fairly inexpensive but requires good siting and its power output varies considerably (which is why it must be mixed with other sources). Biofuels will be an important source of petrol and other liquid fuels, but probably cannot scale up to be a dominant overall energy source due to the stress this would place on land and water resources (see also Chapter 6). Solar power has unlimited potential and prices are declining, but it requires further technical break-throughs to be able to scale up to address the problem at hand. Tidal, hydroelectric and geothermal power are often economical (and geothermal will become more so) but will remain small-scale. Nuclear fusion and other truly novel technologies show no sign of becoming practical in the next few decades.

Predicting the relative costs of these technologies over the next 10 to 20 years is impossible. If costs continue to fall, there are scenarios in which renewable sources could become cheaper than conventional ones even without including their environmental costs. However, most experts doubt this will happen. Renewables will not be widely adopted without some kind of government intervention to supplement private investment, since companies may be reluctant to risk ploughing money into the research and development of technologies which may not succeed or only return very distant profits.

Coal and unconventional oil (shale and tar sands) – fuels even dirtier than coal – are the real climate change battleground. Oil is so useful, and so little remains, that all of it will eventually be used. To avoid massive global warming, however, humans must either leave much of the Earth's bounty of coal and unconventional oil in the ground permanently, or burn it in ways that do not release its carbon into the atmosphere (for example, using carbon capture and storage, or CCS, sometimes known as 'clean coal', discussed in Chapter 4). In July 2010, the United States debated whether to allow or stop a pipeline of much needed oil from Canada, which derived from tar sands – the most carbon-intensive fuel on the planet. It is not encouraging that the Australian State of Victoria is expanding its use of brown coal, the local king of dirty fuels.

One must also look outside the energy sector, to areas such as agriculture, forestry and industry. These sectors are considered in Chapter 4, in the context of their contribution to emissions in Australia and efforts to reduce them.

Many observers are coming to the reluctant conclusion that humanity will not reduce its greenhouse gas emissions, and that a Plan B is needed. Several options, loosely called 'geo-engineering', are indeed available to artificially remove carbon dioxide from the air or otherwise cool the planet.[58] Free-air capture of carbon is technically possible, but is even more uncertain than CCS and currently only works in expensive and material-intensive micro-scale demonstration projects. To capture enough carbon to matter, infrastructure similar to that which currently generates the CO_2 would need to be built. Even with technical advances, it is hard to imagine this ever becoming less expensive or difficult than simply avoiding the emissions in the first place. A more mundane option is to grow and bury biomass, for example as 'biochar'. This would be difficult to scale up to the full size of the problem, and the bio-char would have to be put somewhere where it would not decompose and re-enter the atmosphere even after centuries.

The most plausible geo-engineering option is to make the earth reflect more incoming sunlight. Just changing the proportion reflected from its current 30 per cent to 31 per cent, for example, would have almost the same effect on global temperature as halving the CO_2 concentration. This could be accomplished relatively inexpensively with a fleet of aircraft ejecting a fine spray of (relatively harmless) sulfate in the stratosphere, much like a permanent volcanic eruption.

Before people get too excited, serious ethical and practical problems need to be pointed out. First, cutting down sunlight would not restore local rainfall to its former state in many regions, and might even compound some of the effects of greenhouse gases. Secondly, it would do nothing about acidification. Thirdly, the program to do it would have to continue for centuries; if it ever stopped, most of the global warming that had been suppressed would surge back within a decade. Finally, there is the issue of who (if anyone) has the right to decide to do it. Nonetheless it is likely, for better or for worse, that someone will try it once climate change has really begun to bite.

H Cost analysis of climate change mitigation

The climate change problem cannot be completely solved nor immediately improved. One question then is how much is worth sacrificing now to reduce its severity later? A small branch of economics has developed to answer this very question, in which economists seek trajectories for society that maximise 'intergenerational welfare'. The answer depends on many variables – how much warming will occur per unit of emissions; how expensive the impacts of a given amount of warming will be; how expensive and effective various mitigation approaches are; how the welfare of future generations is valued relative to our own; and how much concern there is for the poor relative to the wealthy. Economists rely on physical science to help with the first two of these questions, and on technical assessments to help with the third; the last two, however, have been largely matters of economic convention, and controversial.

Estimates of the direct economic impact of climate change are difficult to make because only the more obvious effects of a shifting climate can be predicted in advance. Even small changes to biological systems, like the introduction or extinction of a species, can have much broader impacts than one might have expected. Still, economists have attempted to put ranges on the impacts, sector by sector, using a variety of models and analyses of the observed relationships - say, between climate variables and crop yields. These impacts are detailed in Chapter 2 in the context of expected costs to Australia.

There are obviously costs to natural ecosystems and human culture that are hard to measure economically. It may seem inexpensive to relocate people, for example, but if people were happy to leave their cherished homelands there would not be so many people fighting over patches of desert in the Middle East or returning to flood plains or earthquake zones after losing their homes. Such understandable cultural attachments point to a huge hidden cost of relocation. Natural systems likewise presumably have some intrinsic value (apart from any tourism income they may generate), but this value is generally not counted in conventional measures of economic wealth either. When quantifying something is difficult and inherently subjective, as with any unique asset for which no market exists, there is a tendency to ignore it. Those economic analyses that *have* tried to include non-economic costs in some ad-hoc fashion often find them comparable to the economic costs.[59]

The cost figures are also sensitive to some fundamental assumptions. Perhaps the most important one is the value of the so-called 'discount rate'. In effect, the discount rate measures two things: the reduction in value placed on future generations relative to the present, and the rate at which it is expected that people's economic circumstances will continue improving over the next few centuries.[60] Conventionally, values of up to 5 per cent per year or more are sometimes used, which implies either that future generations will be very rich compared to today, or that that there is not much concern now for them. For example, a 5 per cent rate implies that $1 million worth of damage occurring 200 years from now (in today's dollars) is only worth preventing today if it can be done for $57.

The most well known comprehensive review of the economics of climate change to date was undertaken for the British government in 2006, known as *The Stern Review*.[61] This review, which adopted a very low discount rate, predicted that stabilising greenhouse gas levels at acceptable levels of CO_2 (at 500–550 ppmv) would cost around 1 per cent of annual global GDP by 2050 – a relatively small cost compared with the much larger economic harm they calculated would be caused by inaction. Chapter 2 discusses the impacts – and likely costs – of climate change on Australia in particular.

Notes

1 Weart, S, 2004, *The Discovery of Global Warming*, 2nd rev edn, Harvard University Press, Cambridge, p 228; supplemented by <http://www.aip.org/history/climate/index.htm>.

2 Ruddiman, WR, 2001, *Earth's Climate: Past and Future*, Freeman, New York.

3 Lambeck, K, 2002, 'Sea-level Change from Mid-Holocene to Recent Time: An Australian Example with Global Implications' in Mitrovica, JX and Vermeersen, B (eds), 2002, *Glacial Isostatic Adjustment and the Earth System*, American Geophysical Union, Washington DC, pp 33–50.

4 Haub, C. 2002, 'How Many People Have Ever Lived on Earth?', *Population Today*, Population Reference Bureau, Washington, DC, available at <http://www.prb.org/Articles/2002/HowManyPeopleHaveEverLivedonEarth.aspx>.

5 Mann, ME et al, 2009, 'Global Signatures and Dynamical Origins of the Little Ice Age and Medieval Climate Anomaly' 326 *Science* 1256; Cook, ER, Bird, T, Peterson, M, Barbetti, M, Buckley, B, D'Arrigo, R, Francey, R and Tans, P, 1991, 'Climate-change in Tasmania Inferred from a 1089-Year Tree-ring Chronology of Huon Pine' 253 *Science* 1266.

6 Trenberth, KE, Jones, PD, Ambenje, P, Bojariu, R, Easterling, D, Klein Tank, A, Parker, D, Rahimzadeh, F, Renwick, JA, Rusticucci, M, Soden, B and Zhai, P, 2007, 'Observations: Surface and Atmospheric Climate Change' in Intergovernmental Panel on Climate Change (IPCC), 2007, *Climate Change 2007: The Physical Science Basis: Contribution of Working Group I to the Fourth Assessment Report of the Intergovernmental Panel on Climate Change*, Cambridge University Press, Cambridge, ch 3, p 235.

7 Allison, I et al, 2009, *The Copenhagen Diagnosis*, available at <http://www.copenhagendiagnosis.org>.

8 Kiehl, JT and Trenberth, KE, 1997, 'Earth's Annual Global Mean Energy Budget' 78 *Bulletin of the American Meteorological Society* 197. Note that this calculation assumes the Earth absorbs the same percentage of incoming sunlight regardless of temperature.

9 Harries, JE, Brindley, HE, Sagoo, PJ and Bantges, RJ, 2001, 'Increases in Greenhouse Forcing Inferred from the Outgoing Longwave Radiation Spectra of the Earth in 1970 and 1997' 410 *Nature* 355.

10 Weart, above n 1.

11 Le Quéré, C et al, 2009, 'Trends in the Sources and Sinks of Carbon Dioxide' 2 *Nature Geoscience* 1.

12 Berner, RA and Kothavala, Z, 2001, 'GEOCARB III: A Revised Model of Atmospheric CO2 over Phanerozoic Time' 301 *American Journal of Science* 182.

13 Pagani, M, Caldeira, K, Archer, D and Zachos, JC, 2006, 'An Ancient Carbon Mystery' 314 *Science* 1556.

14 Lorius, C, Jouzel, J, Raynaud, D, Hansen, J and Le Treut, H, 1990, 'The Ice-core Record: Climate Sensitivity and Future Greenhouse Warming' 347 *Nature* 139.

15 Jansen, E et al, 'Palaeoclimate: Glacial-Interglacial Variability and Dynamics' in IPCC, above n 6, sect 6.4.1, p 444; Archer, D, Winguth, A, Lea, D and Mahowald, N, 2000, 'What Caused the Glacial/Interglacial Atmospheric pCO(2) Cycles?' 38 *Reviews of Geophysics* 159; Köhler, P, Bintanja, R, Fischer, H, Joos, F, Knutti, R, Lohmann, G and Masson-Delmotte, V, 2010, 'What Caused Earth's Temperature Variations During the Last 800,000 Years? Data-based Evidence on Radiative Forcing and Constraints on Climate Sensitivity' 29 *Quarternary Science Reviews* 129.

16 Mann et al, above n 5.

17 Forster, P and Ramaswamy, V, 2007, 'Changes in Atmospheric Constituents and in Radiative Forcing' in IPCC, above n 6, ch 2, p 129.

18 Yu, H, Kaufman, YJ, Chin, M et al, 2006, 'A Review of Measurement-based Assessments of the Aerosol Direct Radiative Effect and Forcing' 6 *Atmospheric Chemistry and Physics* 613.

19 Hegerl, GC, Zwiers, FW, Braconnot, P, Gillett, NP, Luo, Y, Marengo Orsini, JA, Nicholls, N, Penner, JE and Stott, PA, 'Understanding and Attributing Climate Change' in IPCC, above n 6, sect 9.6, pp 718–26.

20 Royer, DL, Berner, RA and Park, J, 2007, 'Climate Sensitivity Constrained by CO2 Calculations from the Last 420 Million Years' 446 *Nature* 530; Forest, CE, Stone, PH and Sokolov, AP, 2006, 'Estimated PDFs of Climate System Properties Including Natural and Anthropogenic Forcings' 33 *Geophysical Research Letters* L01705.

21 Hegerl et al, above n 19, pp 683–704.

22 Lockwood, M, 2009, 'Solar Change and Climate: An Update in the Light of the Current Exceptional Solar Minimum' 466 *Proceedings of the Royal Society A: Mathematical, Physical & Engineering Sciences* 303.

23 Rahmstorf, S et al, 2007, 'Recent Climate Observations Compared to Projections' 315 *Science* 5810.

24 Doran, PT and Kendall Zimmerman, M, 2009, 'Examining the Scientific Consensus on Climate Change' 90 *Eos, Transactions, American Geophysical Union* 22; Anderegg, WRL, Prall, JW, Harold, J and Schneider, SH, 2010, 'Expert Credibility in Climate Change' 107 *Proceedings of the National Academy of Sciences* 12107.

25 Oreskes, N, 2004, 'Beyond the Ivory Tower: The Scientific Consensus on Climate Change' 306 *Science* 1686.

26 2010, *The Science of Climate Change: Questions and Answers*, Australian Academy of Science, Canberra.

27 Frank, DC, Esper, J, Raible, CC, Buentgen, U, Troutet, V, Stocker, B and Joos, F, 2010, 'Ensemble Reconstruction Constraints on the Global Carbon Cycle Sensitivity to Climate' 463 *Nature* 527.

28 Montenegro, A, Brovkin, V, Eby, M, Archer, D and Weaver, AJ, 2007, 'Long Term Fate of Anthropogenic Carbon' 34 *Geophysical Research Letters* L19707.

29 Solomon, S, Plattner GK et al, 2009, 'Irreversible Climate Change Due to Carbon Dioxide Emissions' 106 *Proceedings of the National Academy of Sciences* 1704.

30 Meinshausen, M et al, 2009, 'Greenhouse-gas Emission Targets for Limiting Global Warming to 2°C' 458 *Nature* 1158.

31 Archer, D, 2005, 'Fate of Fossil Fuel CO2 in Geologic Time' 110 *Journal of Geophysical Research* C09S05.

32 Montenegro et al, above n 28.

33 Sherwood, SC and Huber, M, 2010, 'An Adaptability Limit to Climate Change Due to Heat Stress' 207 *Proceedings of the National Academy of Science* 9552.

34 Meinshausen et al, above n 30.

35 Zachos, J, Pagani, M, Sloan, L, Thomas, E and Billups, K, 2001, 'Trends, Rhythms, and Aberrations in Global Climate 65 Ma to Present' 292 *Science* 686.

36 Ibid; Pearson, PN, Foster, GL and Wade, BS, 2009, 'Atmospheric Carbon Dioxide through the Eocene-Oligocene Climate Transition' 461 *Nature* 1110; Liu, Z, Pagani, M, Zinniker, D, DeConto, R, Huber, M, Brinkuis, H, Shah, SR, Leckie, RM and Pearson, A, 2009, 'Global Cooling During the Eocene-Oligocene Climate Transition' 323 *Science* 1187.

37 Allison et al, above n 7.

38 CSIRO, 2009, *The Science of Tackling Climate Change*, 33.

39 Gilman, EL, Ellison, J, Duke, NC and Field, C, 2008, 'Threats to Mangroves from Climate Change and Adaptation Options: A Review' 89 *Botany* 237.

40 Ridgwell, A and Schmidt, DN, 2010, 'Past Constraints on the Vulnerability of Marine Calcifiers to Massive Carbon Dioxide Release' 3 *Nature Geoscience* 96.

41 De'ath, G, Lough, JM and Fabricius, KE, 2009, 'Declining Coral Calcification on the Great Barrier Reef' 323 *Nature* 116.

42 Allan, RP and Soden, BJ, 2008, 'Atmospheric Warming and the Amplification of Precipitation Extremes' 321 *Science* 1481.

43 Held, IM and Soden, BJ, 2006, 'Robust Responses of the Hydrological Cycle to Global Warming' 19 *Journal of Climate* 5686.

44 Williams, JW, Jackson, ST and Kutzbacht, JE, 2007, 'Projected Distributions of Novel and Disappearing Climates by 2100 AD' 104 *Proceedings of the National Academy of Sciences USA* 5738.

45 Adams, HD, Guardiola-Claramonte, M, Barron-Gafford, GA, Villegas, JC, Breshears, DD, Zou, CB, Troch, PA and Huxman, TE, 2009, 'Temperature Sensitivity of Drought-induced Tree Mortality Portends Increased Regional Die-off under Global-change-type Drought' 106 *Proceedings of the National Academy of Sciences USA* 7063.

46 Friedlingstein, P et al, 2006, 'Climate-carbon Cycle Feedback Analysis: Results from the C4MIP Model Intercomparison' 19 *Journal of Climate* 3337.

47 Battisti, DS and Naylor, RL, 2009, 'Historical Warnings of Future Food Insecurity with Unprecedented Seasonal Heat' 323 *Science* 240.

48 Ibid.

49 Sherwood and Huber, above n 33.

50 Nordhaus, WD, 2005, 'Geography and Macroeconomics: New Data and New Findings' 103 *Proceedings of the National Academy of Sciences USA* 3510; Tol, RSJ, 2009, 'The Economic Effects of Climate Change' 23(2) *Journal of Economic Perspectives* 29.

51 Gupta, S and Tirpak, DA, 2007, 'Policies, Instruments and Co-operative Arrangements' in IPCC, above n 6, ch 13.

52 Hawke, A, 2010, *Review of the Administration of the Home Insulation Program: Report to the Minister for Climate Change and Energy Efficiency*, p 66.

53 United States Environmental Protection Agency (EPA), 2010, *Inventory of U.S. Greenhouse Gas Emissions and Sinks: 1990–2008*, EPA, Washington DC.

54 Commonwealth Treasury, *Australia's Low Pollution Future: The Economics of Climate Change Mitigation*, available at <http://www.treasury.gov.au/lowpollutionfuture/report/html/03_Chapter3.asp>.

55 EPA, 2010, *Global Greenhouse Gas Data*, available at <http://www.epa.gov/climatechange/emissions/globalghg.html>.

56 Pacala, S and Socolow, R, 2004, 'Stabilization Wedges: Solving the Climate Problem for the Next 50 Years with Current Technologies' 305 *Science* 968.

57 Montenegro et al, above n 28.

58 UK Royal Society, 2009, *Geoengineering the Climate: Science, Governance and Uncertainty*, p 84.

59 Weitzman, ML, 2007, 'A Review of the Stern Review on the Economics of Climate Change' 45 *Journal of Economic Literature* 703; Hope, CW, 2006, 'The Marginal Impact of CO2 from PAGE2002: An Integrated Assessment Model Incorporating the IPCC's Five Reasons for Concern' 6 *Integrated Assessment Journal* 19.

60 Tol, RSJ, 2005, 'The Marginal Damage Costs of Carbon Dioxide Emissions: An Assessment of the Uncertainties' 33 *Energy Policy* 2064.

61 Stern, N, 2007, *The Economics of Climate Change: The Stern Review*, Cambridge University Press, Cambridge.

Chapter 2

Climate Change and Australia: A Vulnerable Continent in a Vulnerable Region

Key Issues

- Australia is one of the developed countries that is most vulnerable to climate change, given its high population density in coastal areas, poor soils, variable and dry climate (with scarce water) and extreme weather events. If the economy grows at current rates and emissions are not reduced, Australian temperatures are expected to be 2–3°C above pre-industrial levels by mid-century.
- In Australia, climate change will affect the frequency and intensity of bushfires and severe weather events; sea-level rise; human health, disease and lifestyle; agricultural productivity; energy requirements; and natural ecosystems.
- There will be adverse flow-on impacts on financial services and insurance, trade, tourism, worker productivity and leisure opportunities in hot environments and, more generally, investment, employment and overall economic growth.
- By 2100, unmitigated climate change is predicted to cost annually $50–200 billion to the Australian economy, or $5000–$20,000 per household annually. There will also be less quantifiable costs to Australia's diverse and precious ecosystems.
- Changing rainfall patterns are unpredictable but will substantially affect Australia's experience of climate change. Already dry regions will probably receive less rain, currently temperate zones will become more like inland deserts, droughts will occur more frequently and be more serious, and bushfires are likely to be more common and more intense.

- Rainfall may increase in tropical areas, causing more and heavier cyclones. Generally, rainfall will come in bigger and less frequent bursts, leading to more floods and the need for stronger dams. Precipitation that currently falls as snow will fall as rain, harming alpine biodiversity and tourism in snowfields. A modest increase in the frequency and severity of these weather events would cost hundreds of millions of dollars in direct damage to the built environment annually, and bring greater loss of life.
- Almost all natural systems, but especially coral reefs, forests, alpine systems, wetlands and riverine environments, are sensitive to climate change and will be altered by it.
- The agricultural sector, which represents over one-third of Australia's exports, is expected to be most affected by climate change. The costs of climate change could exceed $1 billion by 2030 and will result from water scarcity, changes in pasture and crop yields, and more pests and diseases. Farmers adapting to new conditions, and some benefits such as fewer frosts, will mitigate some of these effects.
- The Great Barrier Reef will be heavily affected. It will bleach under warm conditions and be damaged by ocean acidification, and the costs of bleaching are conservatively estimated to be over $2 billion per year. Damage to other natural resources will cause further losses.
- 80 per cent of Australia's population lives in coastal areas. Globally, sea-levels are expected to rise by 20–50 centimetres by 2050, which puts $25 billion of fixed assets at direct risk. The greater risk comes from increased storm damage, flooding and coastal erosion.
- The Australian lifestyle – including outdoor labour and sporting events – will be affected by hotter and more humid summers. Tropical diseases are likely to spread further south.
- Global warming could be more serious than predicted, especially by the 22nd century. With 4–6°C of warming, many more ecosystems (including forests) would no longer be viable; tropical regions may become uninhabitable and unsuitable for food production; much of southern Australia would likely be transformed into desert. Eventually, sea-levels would rise by tens of metres, leading to mass migration from coastal areas.

- The regional effects of climate change, especially in countries whose populations are less able to adapt to climate change, will have impacts for Australia's trade, aid, development, and security interests in the Asia Pacific and conservation interests in the Antarctic.

A Introduction

Of all developed countries, Australia is recognised as one of the most vulnerable to climate change. The vast majority of Australians live in coastal areas that are threatened by rising sea-levels. Australia generally has poor soils, a highly variable and typically dry climate, and disastrous extreme weather events (such as bushfires, droughts and floods), all of which will be exacerbated by global warming. Water scarcity has already led to rationing both in major cities and for agricultural use, a situation that could be made worse by climate change. According to the CSIRO and the Australian Bureau of Meteorology, by 2010 average temperatures in Australia had already warmed by almost 1°C since 1950.[1]

By 2050, under a business-as-usual emissions scenario – where the economy grows as usual, and action is not taken to reduce Australia's carbon emissions – annual mean temperatures across Australia are expected to be roughly 2–3°C above their pre-industrial values, with more warming inland than at the coasts. Rainfall patterns will also have changed. Assuming Australia does little to curb emissions, this chapter explores what life may be like in Australia by mid-century, as well as how much it might cost the economy.

Although there can be no definitive evaluation of the precise economic costs of a changing climate, it is possible to take the main categories of direct impact and put sufficient bounds around the problem to understand the magnitude involved. Major categories of direct impact include: fires and severe weather events; sea-level rise; human health, comfort and disease; agricultural productivity; energy requirements; and natural systems. Flowing on from these direct impacts are the resulting impacts on financial services and insurance, trade, tourism, worker productivity and leisure opportunities in hot environments and, more generally, investment, employment and overall economic growth.

Globally, it is estimated that the direct and indirect economic impacts of (unmitigated) climate change by 2100 will include a 5–20 per cent reduction in per capita consumption and a 5–10 per cent reduction in gross domestic product (GDP).[2] In today's terms, and at today's level of GDP, that would equate to an annual cost of $50–200 billion to the Australian economy, or $5000–$20,000 per household annually, by the end of the century. Even if tough action is taken to curb carbon emissions, there will still be costs involved in responding to the level of climate change that is already occurring and will continue to affect Australia for many decades.

The directly measurable economic impact of climate change comes on top of less easily quantifiable cost of the harm to Australia's diverse ecosystems. These systems clearly have value as assets in themselves, over and above their role as tourism drawcards and the cash flow they generate, and even though this value is normally not counted in conventional measures of economic wealth. It is not counted because quantifying it is difficult and subjective, as with any unique asset for which no market exists (for example, how much is the Sistine Chapel worth?). In financial terms, these valuable assets sit on a balance sheet that will be progressively written down as the effects of climate change take hold. To make rational decisions, care must be taken to recognise costs such as these, rather than implicitly assuming that they are zero.

B Changing rainfall patterns?

Much of the impact of climate change on Australia depends on how rainfall patterns change, a matter that remains hazy. Most of Australia lies in a latitude belt that is consistently predicted by climate models to experience less rain in a warmer climate (see Chapter 1) and, not surprisingly, most of the continent is predicted to see less rain. However, clear downward trends in rainfall are not evident so far except in a few regions. The tropical regions of the country will likely become rainier, though this is also uncertain. One reason for the uncertainty is that science still only has a rough idea of what 'normal' is in much of Australia, given that large natural variations in rainfall can persist for decades, and that few rainfall records go back much more than a century.

The predictions of reduced rainfall are thought to be most reliable on the south and west coastal regions, where rain in 2050 could be as little as 60 per cent of typical 20th century values.[3] Southwest Australia has recently experienced a severe drought, which studies have linked at least in part to global warming over and above the normal experiences of drought in the dry Australian continent.[4] Those living in the southwest of Australia will therefore find their climate is becoming more like that of the inland deserts. It is likely that the climates of South Australia and Victoria will also be drier and more like those currently found farther to the north, though probably not as severe as in the southwest. In the southern Murray-Darling Basin, rainfall will also probably be less plentiful than at present,[5] although the drought of the past decade or so in this basin is unlikely to be purely associated with global warming and is not a direct harbinger of the future.

In fact, Australia has always faced an unusual degree of variability in drought conditions. There have been three major droughts in southeast Australia in the past century, each lasting nearly a decade and taking a severe toll on agriculture. It is certain that over the next century more droughts will occur, and would do so even if global warming were somehow prevented. The fickle rainfall patterns here are easily shifted by El Niño, Indian Ocean temperatures and other remote influences. This variability is widely regarded as a prime reason that the aboriginal peoples remained hunter-gatherers until the arrival of Europeans rather than developing agriculture, and it will continue to plague agricultural efforts. The impact of global warming will likely be to make a drought arriving around mid-century somewhat more severe than it would have been, and to reduce the rainfall that arrives during the in-between periods.

For those living in Sydney and elsewhere along the densely populated and relatively moist central east coast, it is still unknown how rainfall might change in the future. Observational studies show that the relatively generous rainfall in this narrow coastal strip is controlled by different processes from those that prevail elsewhere in the country – for example, its rainfall is not affected by El Niño in the way one finds on the western side of the ranges. Therefore, predictions of continental change to this coastal strip cannot be confidently extrapolated. Global climate models cannot resolve the microclimate in this narrow strip, either. For similar reasons, it is uncertain what rainfall Tasmanians will face although recent trends are strongly downward on most of the island.[6]

C Droughts, drenchings and other extremes

Widespread reductions in rainfall, even though they may not be dramatic except in certain regions, will probably leave most Australians coping with more serious drought issues in the future than at present. Increased temperatures will mean that soil moisture drops even in areas where rainfall remains the same. Thus decreases in soil moisture will be worse than one might expect based only on future rain rates. Droughts in most of Australia are forecast to occur 20–50 per cent more frequently by 2050,[7] although, again, northern regions and possibly Tasmania and the east coast may be spared. A mitigating factor is that, in a higher carbon dioxide world, plants will (to varying degrees) not transpire as much moisture into the air, and this will reduce their water demand. This effect is not expected, however, to be sufficient to prevent increases in drought severity.

Another change will have a similar impact: as in the rest of the world, due to a more humid atmosphere, Australian precipitation is expected to come in bigger and less frequent bursts of rain. The intensity of the most extreme rainfall events (in a worst case scenario) could increase by up to 70 per cent in Victoria by 2050 according to one study,[8] with lesser increases expected elsewhere. Whether 50–70 per cent is a reasonable figure remains subject to further investigation, but it is nearly certain that peak rainfall amounts will be significantly heavier. Global observations and theory suggest that peak rainfall would increase by 7 per cent for every 1°C of warming, and possibly up to twice this. This would imply something like 20–30 per cent heavier peak rains in 2050 compared to the 20th century under business-as-usual warming, given similar overall rain amounts. The actual figures should be less, however, in regions where total rain amount decreases significantly.

In the tropical regions, similar increases will apply to rainfall from tropical cyclones, which will become significantly heavier as climate warms. Damaging winds from such storms will also be perhaps 5 per cent stronger by 2050, giving them 10–20 per cent more destructive power.[9] Tropical cyclones will probably occur about as often as they do now, at least according to the latest research, but should be able to reach somewhat farther south. Thus coastal residents living near the Queensland–New South Wales border will be hit by cyclones as often as happens now near the central Queensland coast.

With rain coming much harder, but with little if any increase in total rain, rain in general is expected to become less frequent. This may mean slightly more fair-weather days, good news for outdoors enthusiasts, although it is also possible that drizzle will become more common. Either way, the increasing intermittency of rain delivery will exacerbate problems with water storage, agriculture and soil moisture, and soil erosion. It will also lead to greater flooding problems and more stringent requirements for levees, dams and other water barriers as larger dams will be required to reliably supply the same amount of water.

Another change with some similar impacts is the decrease in the snowpack. Due to rising temperatures, much of the precipitation that in the past has fallen as snow will in the future be falling as rain. By 2050, this is projected to reduce by 50–80 per cent the area receiving snow at least once per year in Australia, while the length of the snow season will contract by 85–96 per cent by 2050.[10] These changes are anticipated to result in the extinction or redistribution of some types of special alpine flora (such as trees, plants and grasses) and fauna (including the mountain pygmy possum), given that the alpine area is already very limited in size.[11] It will also severely reduce the viability of ski resorts and alpine tourism. Even summer activities would also be affected by reduced biodiversity, bushfires and stronger storms.[12]

Bushfires, drought and storms have been the cause of enormous economic and human cost to Australia. The 2009 Victorian bushfires killed 173 people, destroying 3500 homes and other structures, leaving scars in communities and families that will never adequately be quantified.[13] These fires arose from heat and dryness conditions unprecedented in the instrumental record. The Canberra fires of 2003, Victoria's Ash Wednesday fires of 1983 and Tasmania's Black Tuesday fires in 1967 were also devastating. Cyclone Tracy destroyed much of Darwin in 1974, killing 71 people and leaving 41,000 homeless,[14] and property damage from the 1999 Sydney hailstorms was estimated at A$1.7 billion.[15]

Losses from weather-related events in Australia have historically averaged US$719 million per year,[16] but closer to US$1 billion in recent years.[17] The economic cost of the 2011 Queensland floods and tropical Cyclone Yasi which followed soon after (as the strongest cyclone on record in Australia) is estimated to be $5.8 billion.[18] By comparison, this is over twice the estimated annual cost of reducing Australia's carbon emissions to 30 per cent below 1990 levels by 2020.[19]

Science tells us with a high degree of confidence that these extreme weather events will increase in intensity as climate change continues, amplifying the costs as resources and productive labour are diverted to rebuild destroyed assets and infrastructure. By 2050 under a business-as-usual scenario, conditions of very high or extreme fire danger are expected to occur 50–70 per cent more often than in the late 20th century, making bushfires occur more often.[20] These fires will almost certainly be more intense and dangerous.

A modest increase in the frequency and severity of these events would represent hundreds of millions of dollars of direct damage to the built environment annually, as well as additional lives lost. However, impacts may be more than merely modest: one-in-100 year events could potentially become one-in-five or one-in-10 year events,[21] leading to a proportionate scaling up of losses.

Escalating losses of this sort are difficult to insure against, since historical experience becomes an inadequate guide to future risks. This is one reason why the insurance industry has been an early proponent of strong action to combat climate change. The Stern Review estimates that the cost of extreme weather events alone could amount to 0.5 to 1 per cent of global GDP by 2050,[22] and such costs could well affect the viability of insurance companies and ultimately threaten the capacity of individuals and businesses to recover losses in the wake of extreme weather events. As the 2011

Queensland floods demonstrate, recovering from such events can seriously burden even a wealthy country like Australia, which has had to consider new taxes to fund the recovery of public infrastructure and private losses which were not insured (perhaps half of all losses).

D Impacts on ecosystems, forests and plants

Almost all natural systems are threatened in one way or another by climate change, since they have evolved in specific geographies with specific climatic zones and have limits on their adaptability to different conditions. Those likely to feel the impact early include coral reefs, forests, alpine systems, wetlands and riverine environments, all of which are particularly sensitive to climate.

An important knock-on effect of decreasing soil moisture and rising temperatures will be increased stress on forests. The anticipated 2°C or so temperature rise should remain within the tolerance limit of most forests, but may push some marginal forest areas into failing health due to heat stress, and those in regions of significant drying may also fail due to insufficient water.[23] On the other hand, the rising concentrations of carbon dioxide will reduce the water needed by most species and will speed the growth of healthy trees, so that those which are well acclimatised to the warmer climate will grow faster and tree farms will be more productive. This effect will be small in regions of poor soil fertility, because soil nutrients rather than carbon dioxide are the main limiting factor to growth. This means Australian trees will benefit less from carbon dioxide fertilisation than those in most other countries and limit the capacity of Australian forests to absorb more carbon.

A big unknown is how ecosystems will respond to warming, especially milder winters. Many pests are limited by wintertime temperatures, and it is likely that new pest problems will have emerged by 2050. Their impacts are very hard to predict. Already in the early 2000s, huge patches of forests in North America became blighted by pests such as the pine bark beetle, which multiply to huge numbers when there are several mild winters in a row.[24] This used to be very rare but, due to local warming, is becoming typical.

This is a good example of a threshold response that can be surprisingly sensitive to small, but sustained, increases in temperature, and can have an unexpectedly large effect on forest health. It is also possible that other aspects of the ecosystem (for example, predators of the beetle) will eventually respond in a way that pulls the system back toward a relatively healthy new equilibrium.

E Impacts on agriculture and water resources

More than any other sector, agriculture is heavily dependent on the climate and is expected to be most affected by climate change. Our land use has developed around existing climatic zones and rainfall patterns, much of which is expected to be disrupted by human-induced climate change. Agriculture is one of the mainstays of the Australian economy, representing $36 billion of production per year and over one third of Australia's exports, and employing 360,000 people.[25] The costs of climate change in the agriculture sector could thus rise above the hundreds of millions of dollars mark by 2030, and potentially over $1 billion, assuming economic impacts are consistent with past experiences.

Water availability will decrease for much of the populated and productive parts of the country, both urban and rural, with increased dry spells and reduced water supply in dams. Recent droughts have reduced the value of agriculture production by nearly 20 per cent,[26] with the cost of the 2002–03 drought estimated at $6.6 billion. That event reduced Australia's GDP by close to a full percentage point that year. The Murray-Darling Basin, for example, produces over 40 per cent of the gross value of Australia's agricultural production, yet its economic productivity has been predicted to decline by 12 per cent by 2030, 49 per cent by 2050 and 92 per cent by 2100 with unmitigated climate change.[27]

Reduced water flows in the Murray-Darling Basin, changes in pasture and crop growth, increased thermal stress and rising incidences of pests and diseases like the Queensland fruit fly are all expected to have adverse consequences. Livestock productivity is expected to decrease along with the reduction in native pasture growth. Temperate fruit and nut trees will be less successful in

2050, due to the lack of winter chills and the likely expansion of the Queensland fruit fly presence to most of the non-arid parts of the continent. Cattle production will similarly be affected by the expansion of the cattle tick.

The impacts on agriculture are, however, somewhat mixed. Since farmers can adapt rapidly by switching crops or changing use of the land, projections are for the net effect of climate changes by 2050 overall to be comparatively small.[28] Regions currently threatened by frost will become less so; farming will expand into wetter and cooler regions and may become non-viable in regions that are already dry. Sheep farming is expected to be at least as viable as today. Crops such as wheat are expected to become more productive in mass terms (except in very dry areas) in the medium-term due to carbon dioxide fertilisation, but their nitrogen content and nutritive value will correspondingly decrease, and over time increasing temperatures and less moisture will result in an overall decline in yields.[29]

Rangeland farming will likely shift southward and may be as productive as today. Some regions of historical agricultural significance, such as the Barossa grape-growing region, will be less suitable by 2050 due to warmer temperatures, though warming temperatures can be offset by moving up slopes where available and where soil conditions permit. There may also be benefits such as a reduced spread of the light brown apple moth; however, these are expected to be outweighed by the overall costs.[30]

In addition to the agriculture impacts, projected rainfall decreases in the southwest and southeast of the country would affect water availability in the urban areas of Western Australia, South Australia and Victoria. For example, a 1–2°C degree warming is expected to lead to a decrease in Melbourne's water supply by 7–35 per cent.[31] Reduced supply in dams would lead either to stricter water rationing or higher prices to reduce demand from households and businesses. With a higher scarcity value for water and reduced rainfall patterns, the cost of planting trees to mitigate carbon emissions can also be expected to rise, creating a feedback loop of spiralling mitigation costs.

F Impacts on the Great Barrier Reef and tourism

Perhaps the most well known potential victim of heat stress is the Great Barrier Reef. Like most ecosystems, the reef is subject to many human pressures that make it difficult to tell how much of their changes or declines in health so far are due specifically to climate change. In the case of the reef, these additional threats include overfishing, nutrient sources from mainland fertiliser use, and ocean acidification from rising concentrations of dissolved carbon dioxide absorbed from the atmosphere (which results from the same emissions that are causing most of the climate change).

Coral reefs show a clear tendency to bleach under warm conditions: temperatures only 1°C above the long-term summer maxima are sufficient to bleach them. The Great Barrier Reef has warmed by almost 0.5°C over the past century,[32] and experienced eight mass bleaching events from 1979 to 2006, with none recorded before this. In 2002, 60 per cent of the reef was affected by bleaching.[33] By 2050, it is expected that bleaching events will have become annual and widespread, and therefore that corals (which require roughly a decade to recover from a bleaching event) will be dead on most of the reef.[34] Since these impacts do not even include the additional stresses due to ocean acidification caused by rising carbon dioxide, it is unlikely that Australians and tourists of 2050 will still have a lively reef to enjoy.

Projected tourism impacts are among the more directly quantifiable of the economic effects expected as natural resources are harmed by climate change. The Great Barrier Reef attracts over nine million visitors each year, including close to one million from overseas. The tourism revenue it generates is estimated at $4.1 billion, representing a cumulative value over the next 20 years of over $60 billion (taking a discount rate into account).[35] At 2–3°C warming, it is expected that 97 per cent of the Great Barrier Reef will be bleached.[36] If it is assumed, conservatively, that this at least halves the tourism potential of the reef, the economic impact would be expected to be over $2 billion per year in revenue, or a loss of more than $30 billion over the next 20 years. And this ignores the value of the reef to fisheries, or any other intrinsic value.

The profitability of other Australian tourism assets is also likely to be affected by climate change. These include Queensland's tropical rainforests (which neighbour the reef), the alpine ski industry (mentioned earlier), southwest Western Australia (where biodiversity and wine tourism are at risk, and salinity is rising) and the Top End of the Northern Territory (where wetlands and national parks such as Kakadu are at risk from increasing diseases and greater ultraviolet radiation).[37] The unique ecologies of Australia's World Heritage areas could easily be lost.

G Sea-level rise

Reefs will also have to contend with sea-level rise, though this will be far less important for them than for coastal plants and societies. The rates of sea-level rise around Australia will probably not differ greatly on average from those globally, which by 2050 are expected to stand at 20–50 centimetres above those of the late 20th century, and by 2100 between 0.5 and 1.4 metres higher.[38] This will vary by region, as the Australian continent continues to experience slight vertical motions, and fluctuating ocean currents and atmospheric pressure patterns will cause regional variations in sea-level of up to 10 centimetres.

For example, in recent times sea-levels have risen quite rapidly in the north but slowly on the east coast, a pattern driven largely by atmospheric circulation changes. Such changes could reverse and bring large rises to Sydney and Brisbane in the coming decades. Over the long term, sea-level rises will be so large that these regional effects will no longer be very important in a relative sense, but they will still be important in 2050.

Australia is a coastal nation, with 80 per cent of the population living within 50 kilometres of the sea,[39] and 711,000 addresses within three kilometres of the sea. Sea-level rise puts $25 billion of fixed assets potentially at direct risk,[40] although its effects are expected to take place over a time scale of hundreds of years and thus allow some time for adaptation. Medium-term sea-level rise will not cause widespread damage to the population centres of our major cities.

However, increased storm damage, flooding and coastal erosion and inundation is virtually certain to affect low-lying lands and beaches and require the relocation of buildings and people.[41] There is a need for changes to coastal zoning, planning and building laws and policies, and the adjustment of the practices of coastal developers, builders, residents and councils.[42] Unchecked climate change over the long term would lead to rises of many metres, leaving our coastlines and coastal settlements unrecognisably different in future centuries.[43]

H Broader impacts on Australian lifestyle and health

How will people feel in the hot summers and mild winters of 2050? Milder winters will obviously be welcome for many Australians, but the summers will be nearly 2°C hotter than in the late 20th century and humidity will be about 10 per cent higher. In many regions, the humidity increase will have a greater impact on human comfort than the temperature.[44] Summers in Sydney in 2050 will feel roughly like those in Brisbane in the late 20th century, at least in terms of average conditions (Brisbane is only about 2°C warmer than Sydney in the summertime and the relative humidity is about the same in both cities). Brisbane in turn will be even more tropical, and heat stress in the northern tropics will frequently exceed anything experienced on earth in the 20th century.

These changes will certainly mean that those working outdoors can spend less time on the job, especially in the more humid parts of the country. The number of days on which sporting events would be unable to take place due to excessive heat and humidity will increase in southern parts of the country. Heat stress is already responsible for more deaths than more commonly feared disasters such as bushfires and cyclones, and deaths are expected to quadruple by 2050, although this figure depends greatly on adaptation measures.[45] Native animals and livestock will suffer from similar increases in heat stress.

In addition to heat stress and the dangers of extreme weather events, climate change will bring other implications for human health in Australia.[46] As climate zones shift, tropical diseases are expected to spread further out from the tropics. For example, warmer conditions allow the malaria parasite to develop more easily, threatening the reintroduction of malaria to Australia, which was eradicated in 1962 but could spread once again through much of Queensland.[47] Dengue fever is also expected to spread southwards, with the population at risk rising from 0.7 million to between 0.8 and 1.6 million.[48] The direct impact is not only on the health of those afflicted, but the economic flow-on effects such as a decline in labour productivity (as workers require more sick days) and greater public expenditure in the healthcare system.

A direct impact of increased temperatures will be greater demand for cooling in many Australian homes and commercial spaces, along with a reduction in the demand for heating. On balance, the costs are expected to outweigh benefits, with warming of 2–3°C translating to a 3–15 per cent increase in peak electricity demand in Adelaide, Brisbane and Melbourne,[49] offset in part by a decrease in peak demand in Sydney of 1 per cent. This is another feedback loop, since greater energy consumption will lead to accelerating emissions of greenhouse gases from the energy sector unless clean technologies are fully deployed.

Table 2.1 summarises the likely impacts of climate change on Australia, based on the assumptions discussed above. (The worst case scenario is considered in the next section.)

Table 2.1: Overview of Climate Change Impact on Australia

Weather	• Increased rainfall in tropical Australia • Reduced rainfall in the south and west coasts • 20–50 per cent more frequent droughts by 2050 • 20–30 per cent heavier rainfall intensity by 2050 • 10–20 per cent more destructive tropical cyclones • 50–80 per cent reduction in snow area, and 85–96 per cent contraction in snow season, by 2050 • 50–70 per cent more frequent extreme fire danger
Ecosystems and water	• Increased stress on forests and alpine environments • Decreased water availability in many areas • Reduced river flows, including in the Murray-Darling
Agriculture	• Up to $1 billion in losses by 2030 • Specific impacts unpredictable • Decreased livestock productivity • Increased crop pests and diseases • Some positive impacts and shifts in production

Tourism	• Bleaching and destruction of most of the Great Barrier Reef by 2050, with $30 billion lost over the next 20 years • Damage to other protected World Heritage ecosystems • Australian ski-fields severely reduced • Summer activities affected by heat, fires and storms
Sea-levels	• 20–50 centimetres rise by 2050 • Increased storm damage, flooding, coastal erosion • Low-lying lands and beaches affected • $25 billion of fixed assets at direct risk
Lifestyle and health	• Humidity and temperature increases • Outdoor employment and sports more difficult • Heat stress to pets and animals • Spread of tropical diseases • Greater energy demand via air-conditioning
Economy	• 5–10 per cent reduction in GDP by 2100 • Annual cost of $50–200 billion to Australia by 2100 • $5000–$20,000 per household annually by 2100 • Unquantifiable costs to intangible assets

| Worst case scenarios: massive global warming

The impacts described above are expected by mid-century under continued growth in greenhouse gas emissions. With luck, impacts could be somewhat less than this. On the other hand, they could be worse. If stringent controls are put into place to limit emissions, the impacts will be delayed, and in the best case scenario, they may never quite be reached. But what if business-as-usual continues, with no controls put into place? In this case, the world would continue warming, and would probably reach a temperature of 3–5°C above pre-industrial levels by century's end. If, unluckily, the earth system's sensitivity is high, as much as 7°C of warming could occur by then.[50] Even with mid-range 'luck', around 6°C warming would happen by the 22nd-23rd centuries. Thus, it is worth looking at what kind of impacts would be expected under 4°C or 6°C of warming.

Most of the beneficial impacts noted for 2050 would not grow much better with further warming, at least for Australia. By 2050, frost and winter cold would already be relatively minor problems compared to heat stress, for example. However, many of the adverse effects would become dramatically worse under 4°C than 2°C.

For example, while natural forests and other ecosystems (with the exception of reefs and perhaps a few others) can withstand 2°C of warming, few can withstand 4°C. And unfortunately this warming would come on so fast (a single century) that few plant communities would be able to migrate fast enough to keep up; the necessary migration rate is in the order of 5 kilometres per year. This means that many current Australian forests would likely be non-viable in a 4°C warmer world, and would literally go up in smoke and be replaced by savannah or brush. Regions that remain conducive to forest cover (for example, Tasmania) will have changed sufficiently so that current ecosystems are no longer viable, and it is not clear how successful the establishment of new and more heat-tolerant ecosystems would be in such a short period of time or what they would look like. Furthermore, the offsetting gains that will occur in the northern hemisphere due to the 'greening' of Siberia, the Yukon etc will not materialise to any significant extent in the southern hemisphere since the relevant latitudes are covered by the Southern Ocean or Antarctica (where ice sheets would require centuries to melt back and permit forest growth).

Another example where 4°C may exceed a threshold is heat stress. In a 4°C world, with humidity roughly 25 per cent above its 20th century values, it is unclear how suitable for human habitation tropical regions would be, or whether current vegetation would continue to survive the devastating hot spells. It is likely that heat stress on crops, humans and their livestock would combine with other environmental stresses to drive mass emigration from the Northern Territory and northern Queensland as well as Indonesia, India, southeast Asia and other tropical regions, placing additional stresses and costs on other areas of Australia that would be already struggling.

Australian agricultural prospects in a 4°C warmer world are unclear. One should not underestimate the ingenuity of humans in devising new heat-tolerant crops and other technologies, which will probably be essential as many crops grown today would succumb to the heat. The predicted drying of the continent would limit the potential for agriculture in southern parts of the country, but if sufficiently heat-tolerant crops were developed, these could be grown in the water-rich northern regions. Whether human farm workers would be able to withstand the heat there is uncertain, but perhaps robots could do the job or cropping could be done at night. If cheap and efficient desalination were to become a reality, irrigation could be expanded in the more temperate southern regions.

A sustained temperature of 4°C above pre-industrial would, more likely than not, trigger a collapse of the Greenland and/or West Antarctic ice sheets.[51] While this would not occur right away, sea-levels would eventually rise by at least 10 metres, probably over several centuries. Thus a 4°C world would be one in which much of the currently prized real estate in Australia's main cities would be widely recognised as having a limited future. Even though these impacts would come on slowly, the sea-level rise rate by 2100 under business-as-usual would likely reach 10–20 centimetres per decade. At a typical 1:100 coastal slope, this represents a horizontal encroachment rate of 10–20 metres per decade, which would have noticeable impacts on traditional forms of coastal development well within the lifetime of such development. The severity of this would vary regionally depending on how flat the coastal terrain is.

If climate system sensitivity turns out to be moderate or high, and emissions continue unabated, a temperature rise of 6°C is likely some time in the 22nd century. This degree of warming hardly bears thinking about, as all of the impacts mentioned above would be much more severe. Sea-level rise would accelerate to speeds that are difficult to guess but could approach the metre-per-decade rise rates that are evident in the paleoclimate record during periods of rapid deglaciations in the past. Other impacts, however, would be more serious. Any surviving ecosystems would be unrecognisable to us today, and tropical regions would almost surely be so inhospitable as to be unusable for habitation or food production.[52] If forecasts are correct, most of the southern mainland of Australia will have transitioned to desert. The most habitable parts of the country would be Tasmania and the mountainous and coastal regions south of Sydney, where heat stress would still be relatively manageable and water available. But these regions would be unlikely to support a population approaching that of today.

J Impacts on Australia's region: Asia, the Pacific and Antarctica

This chapter has examined what is likely to happen to Australia over the coming decades if carbon emissions are not mitigated and global warming continues unabated. Before the next chapter examines how the international community is attempting to deal with climate change, it is important to consider Australia's experience in the context of its neighbours.

No country is an island; Australia's prospects in a warmer world are also heavily dependent on what happens in its region. Australia's economy is heavily integrated through trade with Asian countries; its foreign aid, development and security policies are oriented towards the Asia Pacific; and Australia is geographically tied to the resources and migration patterns of the region. The Asia-Pacific region will face similar threats to Australia, including higher temperatures and disrupted rainfall patterns, extreme weather events, and adverse impacts on agriculture, ecosystems and biodiversity, human health and livelihoods.

In contrast to Australia, many Asian countries face additional difficulties for their peoples in adapting to the challenges of climate change. High population densities and growth rates, polluted environments, depleted natural resources, ailing public infrastructure and, in some cases, political instability and problems of governance and corruption add to the burden. Food security is a particular problem given the dependence of many people in the region on agriculture. A 2.5–10 per cent decrease in crop yields is predicted in some areas of Asia by 2020, and a 5–30 per cent decrease by 2050.[53] Water scarcity could affect one billion people by 2050.[54] Diseases are likely to increase in countries where public health infrastructure is already struggling.

Sea-level rises and river and coastal erosion and flooding (including from Himalayan glacial melt) will adversely affect millions in low-lying deltas areas such as Bangladesh and the lower Mekong River in Vietnam. Sea-level rises could even threaten the extinction of a Pacific Island country like Tuvalu and seriously undermine the sustainability of life in Kiribati and elsewhere.

Australia's conservation interests in Antarctica will also be affected. Australia maintains scientific bases in Antarctica, claims (disputed) territory there and is an important member of the Antarctic Treaty of 1959 and subsequent environmental agreements. The Antarctic ice sheet is depleting as a result of climate change, resulting in declining krill abundance and adverse impacts on animals (such as certain penguins, seals, sea birds and fish) and ecosystems.[55] Marine and land resources are being affected at Australia's sub-Antarctic islands (such as Heard and McDonald islands). At the same time, given global resources shortages and competition, there is a risk that some countries may seek to exploit natural resources in Antarctica, with profound consequences for the environment and multilateral governance of Antarctica, including Australia's role in it.

The impacts of climate change in the Asia Pacific will impair economic growth, increase poverty and undermine development in countries that are already amongst the world's poorest. The rural and urban poor will be particularly hard hit, from fishers living on the Mekong to vulnerable slum residents in Dhaka. As Chapters 5 and 6 illustrate, some of these impacts will generate the displacement of people and aggravate security problems in the region.

These impacts will also affect Australia's trade, aid and development, and security activities in Asia. Economic downturn in the Asian economies will reduce the need for Australian exports. There will be more demand for aid, development and humanitarian assistance in the region, in addition to the extra burden of finance for climate change adaptation and mitigation. Where climate change aggravates latent security problems in the region (see Chapter 6) or stimulates environmental or economic migration (see Chapter 5), Australia may be called upon to assist through costly humanitarian or security operations, or in resettling the displaced.

Against this background, the next chapter examines how the international community is approaching the problem of climate change, how adequate its response has been and what more needs to be done. The following chapter then turns to Australia's role in the international response and how that connects into Australia's own domestic policy responses.

Notes

1 CSIRO and the Australian Bureau of Meteorology, 2007, *Climate Change in Australia: Technical Report 2007*, CSIRO, Canberra, 6.

2 Stern, N, 2007, *The Economics of Climate Change: The Stern Review*, Cambridge University Press, Cambridge, pp 161–62.

3 Suppiah, R, Hennessy, KJ, Whetton, PH, McInnes, K, Macadam, I, Bathols, J and Ricketts, J, 2007, 'Australian Climate Change Projections Derived from Simulations Performed for the IPCC 4th Assessment Report' 56 *Australian Meteorology Magazine* 131.

4 Nicholls, N, 2010, 'Local and Remote Causes of the Southern Australian Autumn-Winter Rainfall Decline, 1958-2007' 34(6) *Climate Dynamics* 835.

5 CSIRO, 2010, *Climate Variability and Change in South-Eastern Australia: A Synthesis of Findings from Phase 1 of the South Eastern Australian Climate Initiative (SEACI)*, CSIRO, Canberra.

6 Fawcett, RJB, 2010, 'Seasonal Climate Summary Southern Hemisphere (Spring 2009): Rapid Intensification of El Niño, Drier than Average in Northern and Eastern Australia and Warmer than Average Throughout' 60 *Australian Meteorological and Oceanographic Journal* 127.

7 Mpelasoka, F, Hennessy, K, Jones, R and Bates, B, 2008, 'Comparison of Suitable Drought Indices for Climate Change Impacts Assessment over Australia Towards a Resource Management' 28 *International Journal of Climatology* 1283.

8 Whetton, PH, Suppiah, R, McInnes, KL, Hennessy, KJ and Jones, RN, 2002, *Climate Change in Victoria: High-resolution Regional Assessment of Climate Change Impacts*, Consultancy Report for the Department of Natural Resources and Environment, Victoria, CSIRO, May 2002, available at <http://www.water.vic.gov.au/__data/assets/pdf_file/0016/73213/ClimatechangeinVictoria2002.pdf>, p 35.

9 Knutson, TR, McBridge, JL, Chan, J et al, 2010, 'Tropical Cyclones and Climate Change' 3 *Nature Geoscience* 157.

10 Hennessy, KJ, Whetton, PH, Bathols, J, Hutchinson, M, and Sharples, J, 2003, *The Impact of Climate Change on Snow Conditions in Australia*, Consultancy Report for the Victorian Dept of Sustainability and Environment, NSW National Parks and Wildlife Service, Australian Greenhouse Office and the Australian Ski Areas Association, CSIRO Atmospheric Research, available at <http://www.cmar.csiro.au/e-print/open/hennessy_2003a.pdf>.

11 Maunsell Australia Ltd, June 2008, *Impact of Climate Change on Australia's Alpine Areas*, Prepared for the Garnaut Climate Change Review, pp 5, 11, 18.

12 Garnaut, R, 2008, *The Garnaut Climate Change Review: Final Report*, Cambridge University Press, Melbourne, p 135.

13 NSW Parliament, 2009, *Victorian Bushfires: Condolence Motion*, available at <http://www.parliament.nsw.gov.au/prod/PARLMENT/hansart.nsf/V3Key/LA20090313005>.

14 Australian Government Attorney-General's Department, *Emergency Management in Australia*, available at <http://www.ema.gov.au>.

15 Ibid.

16 Intergovernmental Panel on Climate Change (IPCC), 2007, *Climate Change 2007: The Physical Science Basis: Contribution of Working Group I to the Fourth Assessment Report of the Intergovernmental Panel on Climate Change*, Cambridge University Press, Cambridge, ch 11, p 5.

17 Faust, E, 2007, *Climate Change and Insurance Industry: Substantial Risks and Large Opportunities*, Presentation at Greenhouse2007 Conference, Munich Reinsurance Company, Munich.

18 Fraser, A, Nolan R, Queensland Treasury, 21 February 2011, *Statement on Reinsurance*, available at <http://statements.cabinet.qld.gov.au/mms/StatementDisplaySingle.aspx?id=73685>.

19 McKinsey & Company, 2007, *An Australian Cost Curve for Greenhouse Gas Reduction*, McKinsey & Company, Sydney.

20 Hennessy, K, Fitzharris, B, Bates, BC, Harvey, N, Howden, SM, Hughes, L, Salinger J, Warrick, R, 2007, *Australia and New Zealand. Climate Change 2007: Impacts, Adaptation and Vulnerability*, in Parry, ML, Canziani, OF, Palutikof, JP, van der Linden, PJ and Hanson, CE (eds), *Contribution of Working Group II to the Fourth Assessment Report of the Intergovernmental Panel on Climate Change*, Cambridge University Press, Cambridge, para 13.

21 Parkinson, M, 2009, *The Economics of Climate Change and its Policy Implications*, Opening Presentation to the G-20 Workshop on the Economic and Financial Implications of Climate Change, Department of Climate Change, Sydney.

22 Stern, above n 2, p 8.

23 Allen, CD, Macalady, AK, Chenchouni, H et al, 2010, 'A Global Overview of Drought and Heat-induced Tree Mortality Reveals Emerging Climate Change Risks for Forests' 259 *Forest Ecology and Management* 660.

24 Kurz, WA, Dymond, CC, Stinson, G, Rampley, GJ, Neilson, ET, Carroll, AL, Ebata, T and Safranyik, L, 2008, 'Mountain Pine Beetle and Forest Carbon Feedback to Climate Change' 452 *Nature* 987.

25 Department of Climate Change and Energy Efficiency, last updated 2009, *Agriculture*, available at <http://www.climatechange.gov.au/climate-change/impacts/agriculture.aspx>.

26 Ibid.

27 Garnaut, above n 12, pp 129–30.

28 Hennessy et al, above n 20, para 13.

29 Garnaut, above n 12, p 132.

30 Preston, BL and Jones, RN, 2006, *Climate Change Impacts on Australia and the Benefits of Early Action to Reduce Global Greenhouse Gas Emissions*, Consultancy report for the Australian Business Roundtable on Climate Change, available at <http://www.csiro.au/files/files/p6fy.pdf> (accessed 15 March 2011), p 23.

31 Howe, C, Jones, RN, Maheepala, S and Rhodes, B, 2005, *Implications of Climate Change for Melbourne's Water Resources*, Melbourne Water, Melbourne, cited in Preston and Jones, above n 30, p 26.

32 Garnaut, above n 12, p 143.

33 Ibid.

34 Jones, RN, 2004, 'Managing Climate Change Risks' in Agrawala, S and Corfee-Morlot, J (eds), 2004, *The Benefits of Climate Change Policies: Analytical and Framework Issues*, OECD, Paris, p 249, cited in CSIRO, above n 5.

35 Bailey, G et al, 2003, Assessment of Tourism Activity in the Great Barrier Reef Marine Park Region: A Report by the Bureau of Tourism Research for the Great Barrier Reef Marine Park Authority, Bureau of Tourism Research, Canberra.

36 Jones, above n 34.

37 Garnaut, above n 12, p 134.

38 Rahmstorf, S, 2007, 'A Semi-Empirical Approach to Projecting Future Sea-Level Rise' 315 *Science* 368.

39 Garnaut, above n 12, p 137.

40 Wong, P, November 2008, 'The Economics of Climate Change' 78 *Intheblack* 72.

41 Garnaut, above n 12, p 138; see United Nations Environment Programme (UNEP), 2007, 'Global Outlook for Ice and Snow', Global Environment Outlook Series, available at <http://maps.grida.no/go/collection/global-outlook-for-ice-and-snow>.

42 See, eg, McDonald, J, 'The Adaptation Imperative: Managing the Legal Risks of Climate Change Impacts' in Bonyhady, T and Christoff, P (eds), 2007, *Climate Law in Australia*, Federation Press, Sydney, p 124; Bonyhady, T, 'Swimming in the Streets: The Beginnings of Planning for Sea Level Rise' in Bonyhady, T, Macintosh, A and McDonald, J (eds), 2010, *Adaptation to Climate Change: Law and Policy*, Federation Press, Sydney, p 80.

43 Overpeck, JT, Otto-Bliesner, BL, Miller, GH, Muhs, DR, Alley, RB and Kiehl, JT, 2006, 'Paleoclimatic Evidence for Future Ice-sheet Instability and Rapid Sea-level Rise' 311 *Science* 1747.

44 Willett, KM and Sherwood, S, 2012, 'Exceedance of Heat Index Thresholds for 15 Regions under a Warming Climate using the Wet Bulb Globe Temperature' 32(2) *International Journal of Climatology* 161, available at <http://onlinelibrary.wiley.com/doi/10.1002/joc.2257/abstract>.

45 McMichael, A, Woodruff, R, Whetton, P, Hennessy, K, Nicholls, N, Hales, S, Woodward, A and Kjellstrom, T, 2003, *Human Health and Climate Change in Oceania: A Risk Assessment*, Commonwealth Department of Health and Ageing, available at <http://www.health.gov.au/internet/main/publishing.nsf/content/.../env_climate.pdf>.

46 See generally, Garnaut, above n 12, pp 139–41.

47 Perry, M, 2008, 'The Sting of Climate Change', *Reuters*, Sydney, available at <http://www.reuters.com/article/idUSTRE4AJ2RQ20081120>.

48 Ford, M, Gunasekera, D, Tulloh, C, Kim, Y, 2007, *Economic Impacts of Climate Change in Australia: Framework and Analysis,* ABARE Greenhouse 2007 Presentation.

49 Howden, SM and Crimp, S, 2001, (2001) 'Effect of Climate and Climate Change on Electricity Demand in Australia' in International Congress on Modelling and Simulation, 2001, *Integrating Models for Natural Resources Management Across Disciplines: Issues and Scales*, ANU, Canberra; Ghassemi, F, Whetton, P, Little, R and Littleboy, M (eds), *Modelling and Simulation Society of Australia and New Zealand*, MSSANZ, Canberra, pp 655–60, cited in Preston and Jones, above n 30, p 29.

50 Sokolov, AP et al, 2009, 'Probabilistic Forecast for 21st Century Climate Based on Uncertainties in Emissions (Without Policy) and Climate Parameters' 22 *Journal of Climate* 5174.

51 Stone, EJ, Lunt, DJ, Rutt, IC and Hanne, E, 2010, 'Investigating the Sensitivity of Numerical Model Simulations of the Modern State of the Greenland Ice-sheet and its Future Response to Climate Change' 4 *Cryosphere* 397.

52 Sherwood, SC and Huber, M, 'An Adaptability Limit to Global Warming Due to heat Stress' 107 *Proceedings of the National Academy of Sciences* 9552.

53 IPCC, above n 16, ch 10: Asia, p 471.

54 Ibid.

55 IPCC, 2004, *Working Group Report*, Cambridge University Press, Cambridge, ch 15: Polar Regions (Arctic and Antarctic), pp 655, 660.

Chapter 3

Global Climate Law and Politics: From Rio to Cancún and Beyond

Key Issues

- All countries contribute to and are affected by climate change. The problem can only be fully addressed by a binding, global legal regime. The foundations of such a regime are the United Nations Framework Convention on Climate Change (UNFCCC) of 1992, the Kyoto Protocol of 1997, the Copenhagen Accord of 2009 and the 2010 Cancún Agreements.
- The international community has agreed on the need to avoid global average temperature from rising more than 2°C above pre-industrial levels. There is also broad consensus on the need to reduce emissions *immediately*.
- More contentious is how to distribute emission cuts fairly among nations. It is accepted that developed countries should take the most responsibility, and that emissions should be allocated roughly on a per capita basis. Australia has argued that it deserves concessions given its economic reliance on fossil fuels.
- Global policy choices about climate change should be turned into binding rules of international law. Successful law-making of this kind has happened in other areas of global environmental protection since the 1970s, regarding atmospheric pollution and acid rain, marine pollution, biodiversity conservation and ozone depletion. Early action is needed on environmental problems.
- At the UNFCCC negotiations, Australia and many European countries lobbied for binding emissions targets, but these were vetoed by United States and oil-producing countries. Ultimately, the UNFCCC provided principles for climate change mitigation (including protecting the environment for future generations, attention to the needs of developing countries, and the precautionary principle), rather than a comprehensive legal regime.

- The subsequent Kyoto Protocol imposes binding emissions reduction targets on developing countries, and requires them to provide assistance to developing countries in making their economies less carbon-intensive. The Protocol establishes three market-based 'flexible mechanisms' (joint implementation of reduction projects by two or more industrialised parties; the Clean Development Mechanism; and emissions trading) to achieve its objectives. Its innovative compliance procedure allows countries in breach of the treaty to be suspended from certain benefits or otherwise penalised.
- The Kyoto Protocol was only a first step in combating climate change, as its targets were modest. Many countries, including Australia, advocated replacing the Kyoto system with a new treaty identifying binding emissions reductions.
- The 2006 Asia Pacific Partnership on Clean Development and Climate involved countries that produce more than half of the world's carbon emissions. It is a non-binding arrangement which lacks benchmarks or timeframes, instead placing faith in technological innovation, voluntary programs and government subsidies. Some of its terms may in fact breach the UNFCCC.
- At the rancorous 2009 Copenhagen conference, the traditional split between developed and developing countries was replaced by a new division between larger, wealthier developing countries and the poorest developing countries. The Copenhagen Accord is limited in its utility because it is non-binding and lacks the normative power of even previous United Nations conference declarations.
- The Copenhagen Accord still has some positive aspects, including bringing together the mitigation commitments of developed and developing countries; putting a dollar figure on contributions by developed countries to developing countries to meet adaptation and mitigation needs; addressing emissions from deforestation for the first time; and an innovative Technology Mechanism.
- Most countries recognise the 2010 Cancún Agreements as representing global consensus on climate change better than the Copenhagen Accord. However, the 2010 Agreements did not include an agreement on a second commitment period for the Kyoto Protocol, which expires in 2012.

- Given the failure of the much-hyped Copenhagen and Cancún conferences to produce a strong post-Kyoto agreement, some critics doubt that the UNFCCC process will ever provide an effective solution. Instead, countries may increasingly rely on complementary, sector-specific legal regimes on marine pollution, biological diversity, world heritage sites and trade law. Given the urgency of reducing emissions, the challenge of reaching an ideal international response should not be used as an excuse for inaction.
- Litigation in domestic and international courts and complaints bodies has been an ad hoc legal means of combating climate change. 'Progressive' proceedings seek to enforce emissions cuts against a country. 'Regressive' proceedings may prevent a country from implementing policies to reduce emissions if these interfere with other rules (such as on trade liberalisation). 'Administrative' proceedings under the compliance regime of the Kyoto Protocol may ensure that states adhere to the Protocol's obligations.

A Introduction

In his landmark 1968 essay, 'The Tragedy of the Commons', Garrett Hardin vividly showed how common resources such as open pastures will be overexploited unless there are clear rules setting sustainable limits on access to them. Without regulation it makes perfect sense for a herdsman to add animals to his flock, because the environmental costs of doing so are spread across all resource users and, besides, there is no point in being virtuous when there is no guarantee that others will also do the right thing.

Climate change is a classic tragedy of the commons in that the atmosphere is open to carbon wastes from all humanity, and the problem can only be addressed by an international legal regime binding all of the approximately 200 nations that make up the international community. As Ross Garnaut, the Australian government's chief adviser on climate change, puts it: 'Individual countries will not on their own undertake adequate mitigation, since each country is better off – from a narrow, national point of view – the more it can free ride on the efforts of others'.[1]

The need for collective and coordinated action, backed up by binding international legal rules, is a major, but not insuperable, hurdle to global action on climate change. Important steps have already been taken in developing an international legal framework, even though there are many details that remain to be worked through. The foundations are provided in the United Nations Framework Convention on Climate Change (UNFCCC), signed at the Rio Earth Summit in 1992, the Kyoto Protocol agreed five years later in Japan and the Copenhagen Accord of December 2009. This chapter evaluates these key texts, and looks forward to what still needs to be achieved if we are to fashion a truly effective international law of climate change.

B Global policy choices

Chapter 1 showed how climate science, synthesised by the Intergovernmental Panel on Climate Change (IPCC), provides policy-makers with a wealth of information on the causes and effects of global warming, and what can be done to abate emissions and adapt to climatic changes. There is no precedent in human history for such extensive and authoritative scientific analysis of a global environmental problem. On the basis of the IPCC's assessments, three main policy choices now need to be made by Australia and the international community as a whole.

The first and most important is what atmospheric concentration of greenhouse gases the world is comfortable with. This choice about where the global thermostat is to be set is a question of intergenerational equity: what kind of climate should the present generation leave for those generations to come? The science tells us that the concentration cannot exceed 450 ppmv carbon dioxide equivalent (CO_2e – ppmv stands for parts per million by volume and is a measure of the concentration of a gas) if we are to have a 50 per cent chance of avoiding the global average temperature from rising more than 2°C above pre-industrial levels.[2] The 2°C limit was adopted by the European Union in 1996 (and reaffirmed in 2005), but it was not until the Copenhagen Accord in December 2009 that it attracted general international support.

However, the world may have to aim for concentrations as low at 350 ppmv CO_2e, with a corresponding temperature increase of 1°C, if we are to prevent a number of catastrophic impacts such as the inundation of many small islands in the Pacific and the Caribbean, and the loss of most coral reefs, including great swathes of the Great Barrier Reef. For this reason the Alliance of Small Island States has for some time advocated a global climate deal that would prevent temperatures from rising more than 1.5°C.[3] As noted in Chapter 1, based on current emissions, the world is heading for increases beyond 2°C unless there are rapid efforts to curb global emissions.

The second choice, flowing from the first, is when emissions cuts are to be made to reach the concentration and temperature target. In this respect increasing carbon concentrations in the atmosphere can be likened to water filling a bathtub. The imperative is to keep within safe limits, which in the case of a bathtub means turning off the water before it spills over on to the bathroom floor. This can be done slowly by gradually turning the faucet to restrict the flow of water or the spout can be left to gush until the very last moment and then shut off. As we saw in Chapter 2, when it comes to controlling carbon emissions the most prudent approach is to reduce emissions early, gradually increasing our efforts over time until carbon flows from humanity once again are within the capacity of the earth to absorb them. Unfortunately, there is a natural human tendency to leave matters to the last moment. This is precisely the wrong thing to do in the case of climate change as it could become completely unstoppable and 'shutting off the faucet' may take a very long time indeed. Besides which, nations that wait will miss out on the economic benefits that will flow to states that develop clean energy industries that will be in great demand when a price is placed on carbon.[4]

The third international climate policy decision is the most difficult – which is how to distribute emissions reductions fairly among all the nations that make up the international community. Chart 3.1 (*over page*) shows the countries with the highest carbon emissions worldwide as a percentage of the global total.

In this respect two principles for allocating emissions reductions have assumed prominence. The first of these is that developed

Chart 3.1: Share of carbon dioxide emissions by country, 2008 (%)

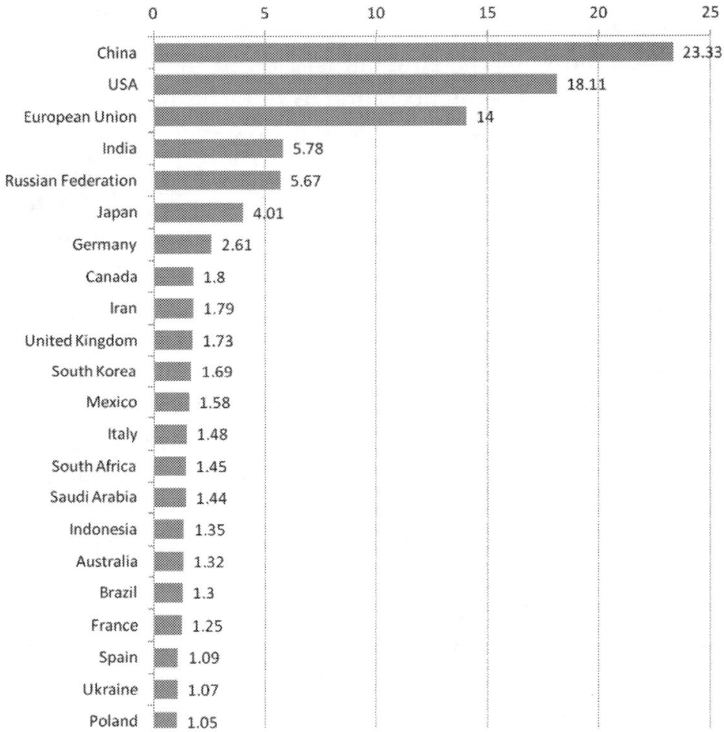

Source: United Nations Statistics Division, 'Millennium Development Goal Indicators', 2011, available at <http://mdgs.un.org/unsd/mdg/SeriesDetail.aspx?srid=749&crid=>.

states should take the lead in reducing emissions because most of the carbon in the atmosphere was put there by them as they industrialised. It is estimated that, between 1850 and 2002, developed countries were responsible for 76 per cent of cumulative CO_2 emissions.[5] Those figures are even starker when it is also recognised that the developed countries house a smaller share of the world's population than the developing countries.

A second and related principle is that emissions allocations between countries should ultimately be on a per capita basis, so that sometime in the future the emissions from a resident of Canberra would equal those of a resident of Kolkata. This is the notion of 'contraction and convergence' pioneered by Aubrey Meyer at the

Chart 3.2: Changes in emissions by country (%), 1990–2004

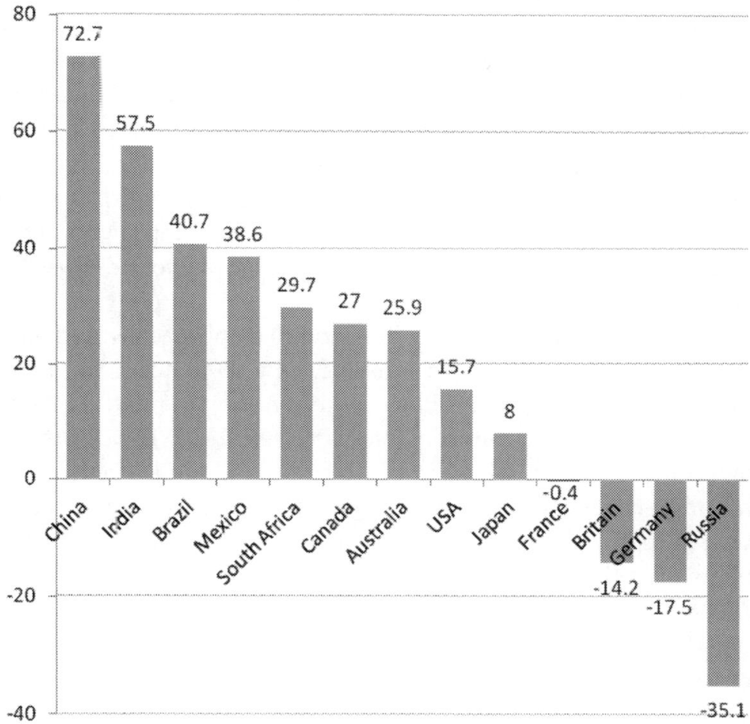

Source: Gabrielle Walker and Sir David King, 2008, The Hot Topic: How to Tackle Global Warming and Still Keep the Lights On, Bloomsbury, London, pp 199–231.

Global Commons Institute, and involves setting a global emissions target and then contracting national emissions over time until the emissions of every person on the planet converges.[6]

Neither of these principles should be understood to excuse developing countries from reducing their emissions. While developing countries currently still emit less carbon than developed countries, the rate of emissions from developing countries has grown much more rapidly in recent years than that from developed countries, and their total share of emissions is growing accordingly, as Chart 3.2 shows.

Reducing emissions in developing countries is increasingly urgent as their economies and carbon intensity grow, but the challenge is ensuring that it is done in a way which also allows their legitimate aspirations for economic development to be met.

Translating these fair and commonsense principles into binding legal obligations has proven difficult as developed and developing states remain divided on how to share the burden of emissions reductions, and whether particular national circumstances should be taken into account. The Garnaut Review (discussed in Chapter 2) adopted a modified contraction and convergence approach, modified in the sense that it recommended that developing countries be given some 'headroom' before having to move to the emissions convergence line.[7]

If aiming at 450 ppmv CO_2e, the modified approach would mean Australia reducing per capita emissions by 95 per cent from 2000 levels by 2050, whereas India would be entitled to raise per capita emissions by 22 per cent from 2000 levels by 2050. The Australian government has not taken up this recommendation. The headline 2050 target of an 80 per cent reduction on 2000 levels is clearly below what is required, as are the mid-term targets. Australian governments of both political persuasions have argued that Australia is a special case in that the economy is heavily dependent on fossil fuels for the production of electricity, and for road, rail and air transport across our expansive continent. On this basis, Australia has adopted 'comparable effort' approach rather than 'contraction and convergence'.

However, in the longer term the Garnaut Review notes that if Australia continues to argue that it is a special case then other nations will want to do the same, and this special pleading will only undermine the environmental effectiveness of an international climate change agreement.[8] The policy choices open to Australia in responding to climate change are considered in the next chapter, after this chapter first sets the scene by investigating the key components of the international framework.

C International law of the atmosphere

It is to international law that we must look to transform global policy choices about climate change into binding rules. International law is the system of norms and principles that has developed over many centuries to bring order and stability to the relations between nation states. International law was traditionally concerned with issues such as the declaration and conduct of wars, the conclusion of treaties and the protection of diplomats. But in the 20th and 21st centuries international law has expanded tremendously and now addresses a host of new challenges such as safeguarding human rights, promoting free trade and protecting the global environment. Understanding current global regulatory attempts to respond to climate change first requires an appreciation of the international community's approach to dealing with environmental challenges generally.

1 International law of the environment

States became aware of environmental issues spanning international borders in the late 19th century, as natural resources such as fisheries and rivers came under increasing pressure and industrial pollution began to cause problems such as acid rain. However, it was only from the 1970s onwards with the advent of environmental movements in many northern states that a more comprehensive global approach began to be taken on environmental protection. At the Stockholm Conference on the Human Environment in 1972 states adopted a non-binding instrument that set out basic principles of global environmental protection that remain relevant to this day. However, international environmental law now encompasses not only familiar general concepts such as 'sustainable development' and the 'precautionary principle', but also hundreds of detailed treaties dealing with all manner of environmental issues, from protecting world cultural and natural heritage, to the prevention of marine pollution, the conservation of biodiversity and the regulation of trade in endangered species. The protection of the atmosphere has also been a particular focus of international law relating to the environment and it is in this context that rules relating to climate change sit.

2 Atmospheric pollution and acid rain

For many decades problems of atmospheric pollution have been high on the international legal agenda. The problem receiving the earliest attention was acid rain that is caused when sulphur dioxide (SO_2) emitted by industry reacts with water in the atmosphere and returns to the ground as sulphuric acid.

International regulation of acid rain stretches back to the *Trail Smelter Case* in the 1930s.[9] That case concerned the emission of SO_2 and other pollutants from a lead and zinc smelter at Trail in British Columbia, Canada, which drifted down the Columbia River Valley and into the United States, where damage was caused to crops in Washington State. Canada was held responsible for a major proportion of this damage by an international tribunal established by Canada and the United States, on the basis of the international legal principle that no state has the right to use, or permit the use, of its territory in such a way as to cause serious injury by pollution in the territory of another state.[10] This cardinal rule of international environmental law subsequently found expression in Principle 21 of the 1972 Declaration of the United Nations Conference on the Human Environment[11] (and was substantially repeated in Principle 2 of the 1992 United Nations Declaration on Environment and Development[12]). The rule is often cited as a potential basis for a claim by states that are clear victims of climate change (such as the Pacific microstate Tuvalu which is threatened by rising sea-levels) against major emitting states (such as the United States and China).[13]

In the Northern Hemisphere a sophisticated legal regime now regulates the transboundary air pollution that causes acid rain and localised ozone pollution, under the 1979 Convention on Long-Range Transboundary Air Pollution. This convention, which was negotiated within the United Nations Economic Commission for Europe, has Pan-European and North American membership. It is a so-called framework convention in that it provides a structure which is supplemented by eight implementing protocols, including several instruments that set specific targets for reducing the acidification effects of atmospheric pollution from sulphur and nitrogen oxide emissions. It is an example of a strong regional agreement on atmospheric pollution and provides an important example of what can be achieved through international cooperation.

3 Ozone depletion

A more recent and wide-ranging problem of atmospheric pollution addressed by international law is ozone depletion. In the late 1970s scientists discovered a gaping hole in the ozone layer that protects the earth from dangerous ultraviolet light that not only causes skin cancer in humans but also damages a range of plants and animals, including krill, the shrimp-like marine invertebrate at the base of the oceanic food chain. Clear scientific evidence and widespread public concern led in a very short timeframe to negotiations on a global regime to protect the ozone layer by phasing out the production and consumption of ozone-depleting chlorofluorocarbons (CFCs) and hydrochlorofluorocarbons (HCFCs), the synthetic substances invented by the Dupont corporation in the 1920s and used in all manner of everyday products from hairspray to refrigerators. Ozone-depleting substances also have a very high global warming potential (some CFCs have a warming potential up to 10,000 times that of CO_2) so their elimination is also important in combating climate change.

Having taken strong action at home to reduce emissions of ozone-depleting substances, the United States led international negotiations towards a regime that would have to encompass all states if it were to be truly effective. As in the negotiations on climate change, ozone depletion could only be arrested if all free riders were brought on board. In 1985 agreement was reached by over 130 countries on the Vienna Convention for the Protection of the Ozone Layer. This was a framework treaty setting out general principles and few concrete obligations. However, new scientific evidence confirming the seriousness of ozone depletion led to a further, implementing agreement, the 1987 Montreal Protocol on Substances that Deplete the Ozone Layer, which committed parties to reducing the production and consumption of ozone-depleting substances by 50 per cent on 1986 levels by 1998. Tougher targets were later agreed through the London (1990), Copenhagen (1992), Vienna (1995) Montreal (1997) and Beijing (1999) amendments.

Although Australia is responsible for less than 1 per cent of emissions of ozone-depleting substances, Australia has been a strong supporter of the ozone regime given the special threat that ozone depletion poses to nations in the southern hemisphere where ozone thinning is most pronounced. Australia has ratified the Vienna Convention, Montreal Protocol and all of the amendments agreed to date, and has implemented its international obligations under the regime through the *Ozone Protection and Synthetic Greenhouse Gas Management Act 1989* (Cth).

The ozone regime has been widely regarded as a success story of international cooperation, proving that a precautionary approach can and should be taken to avoid an environmental threat even when this involves significant economic costs. One of the important features of the regime is the way in which developing nations have been brought within its binding rules. Poorer countries were initially not bound to phase out the production and use of CFCs and HCFCs, but following amendments to the Montreal Protocol agreed in 1990, India, China and other states took on these obligations and were compensated for so doing by developed economies. This was only possible following the establishment in 1991 of the world's first financial mechanism under an environmental treaty, the Multilateral Fund for the Implementation of the Montreal Protocol, through which developed states agreed to assist developing country parties comply with control measures.

The ozone regime is on track towards its goal of phasing out emissions of ozone-depleting substances from developed and developing countries by 2030. Between 1986 and 2007 the parties to the Montreal Protocol achieved a 97 per cent reduction in the consumption of ozone-depleting substances.[14] The regime appears to have been successful in stabilising atmospheric concentrations of ozone-depleting substances, such that the ozone layer has begun to recover and may be restored to full health by the end of the century.[15]

4 Climate change

The experience with the ozone regime emphasises the need to confront global environmental problems with early action, before the damage is so great that effective policy choices are constrained, or simply non-existent. Despite the apparent similarities between ozone depletion and climate change as atmospheric pollution problems, the politics surrounding the latter have been far more fraught. The key difference is that the greenhouse problem results from emissions that are more widely distributed than ozone-depleting substances (which were produced in a relatively small number of factories, and could be replaced with alternatives relatively easily). Moreover, the initial refusal by the chemical industry to address the ozone problem met with significant consumer pressure, particularly in the United States where there were boycotts of products such as spray cans containing CFCs. There were also threats that producers such as Dupont would be subject to litigation for contributing, among other things, to increased rates of skin cancer.

In contrast to the ozone problem, consumers in most countries are highly dependent on the energy sources provided to them by public utilities or private corporations, which require long-term, large scale investments (such as in coal-fired power stations). Those choices cannot be so readily influenced by consumer pressure, even if consumers are increasingly asking for change (consider the growing demand for renewable energy and electric cars). Powerful commercial interests, including fossil fuel companies, have also played a role in opposing or delaying change. In addition, unlike with the ozone problem, litigation to challenge carbon emissions is fraught with technical difficulties (see further below) and unlikely to trigger changes on the scale needed to confront the problem.

In 1988 the UN General Assembly recognised climate change as a 'common concern of mankind' and initiated a process of scientific analysis and international attention that led to the conclusion of the UNFCCC and the Kyoto Protocol. Negotiations on the UNFCCC were launched in 1990, and proved not only politically contentious but also logistically difficult, as the international conference took place in a number of sessions in Washington DC, New York, Geneva and Nairobi.[16]

Australia and many other states, including those of the European Community, argued for binding emissions targets in the UNFCCC. However, the United States and oil-producing countries effectively vetoed the inclusion of targets and timetables. A compromise was reached in the carefully drafted text, which committed industrialised countries to returning their emissions to 1990 levels by 2000.[17] Hence the main function of the UNFCCC is not to spell out concrete action but rather to provide a principled foundation for a comprehensive international regime built through the adoption of specific implementing protocols, such as the 1997 Kyoto Protocol.

The overarching goal of the climate regime is set out in Article 2 of the UNFCCC – the 'stabilization of greenhouse gas concentrations in the atmosphere at a level that would prevent dangerous anthropogenic interference with the climate system'. The provision goes on to state that this level should be achieved within a timeframe that will permit ecosystems to adapt naturally, prevent threats to food production and ensure that economic development can proceed sustainably. Significantly, Article 2 does not specify a safe atmospheric concentration of CO_2e, which means that the international community must have regard to the latest climate science when tailoring emissions reductions policies to meet the goal of the international climate regime. There is ongoing debate around this issue, as we saw in Chapter 1.

In pursuing the UNFCCC's objective the parties are to be guided by several principles. The first of these is that the climate system is to be protected for the benefit of present and future generations, on the basis of equity and according to common but differentiated responsibilities and respective capabilities.[18] To this end developed countries must take the lead in mitigating climate change and assisting all peoples in adapting to change that cannot be averted.

Second, particular attention must be paid to the needs and special circumstances of developing countries, especially those states that are vulnerable to climate change such as small island developing states in the Pacific and elsewhere.

Third, the UNFCCC incorporates the precautionary principle so that full scientific certainty should not be used as an excuse for failing to take measures to anticipate, prevent or minimise the causes of climate change. This principle is important because, while climate science tells us with great certainty that greenhouse gases from human activities are changing the climate, there are uncertainties as to when and by how much the climate will change and how resilient natural and social systems will be. The precautionary principle recognises the high stakes involved in inaction.

The UNFCCC divides countries into two categories: industrialised nations, including economies in transition such as eastern European states, and non-industrialised nations, that is the developing states that make up the majority of the international community. The industrialised group of countries are listed in Annex I of the UNFCCC and so are usually known as Annex I countries. The UNFCCC imposes some obligations on all parties such as duties to monitor and report their greenhouse gas emissions, and develop programs to mitigate climate change. But it is the Annex I states that are given special duties on the 'basis of equity' and in accordance with the principle of 'common but differentiated responsibilities and capabilities' to 'take the lead in combating climate change and the adverse effects thereof'.

5 1997 Kyoto Protocol

The UNFCCC is one of the most widely ratified treaties in history, with 192 countries participating in it – virtually the whole international community. Yet almost as soon as the ink was dry on the UNFCCC text it was clear that much more needed to be done to control greenhouse gas emissions. The Kyoto Protocol was agreed in 1997 and was the product of three years of negotiations that began at the First Conference of the Parties to the UNFCCC in Berlin in 1995.

The Kyoto Protocol operationalises the central bargain agreed in the UNFCCC, namely that developed countries would take on legally binding emissions reduction or limitation targets, and in providing technological and other assistance to developing countries to help them develop less carbon-intensive economies. As with the Montreal Protocol the expectation was that, once developed countries had demonstrated real progress in reducing emissions in several commitment periods, developing countries would begin to assume similar obligations.

The fundamental objective of the Kyoto Protocol is set out in Article 3, which states that the Annex I parties must

> Individually or jointly, ensure that their aggregate anthropogenic carbon dioxide equivalent emissions of the greenhouse gases listed in Annex A do not exceed their assigned amounts,

> calculated pursuant to their quantified emission limitation and reduction commitments inscribed in Annex B ... with a view to reducing their overall emissions of such gases by at least 5 per cent below 1990 levels in the commitment period 2008-2012.

Five year commitment periods allow parties to average their emissions over a number of years and smooth the effects of one-off events that could lead to unexpected spikes in emissions. In Annex A to the Kyoto Protocol six greenhouse gases are listed: carbon dioxide, methane, nitrous oxide, hydrofluorocarbons, perfluorocarbons and sulphur hexafluoride. This is by no means an exhaustive list, but it captures the most important of the greenhouse gases produced from human activities. What Kyoto imposes are limits on the emissions of carbon dioxide equivalent - and so it is up to individual parties how they will meet their targets by abating emissions of one or more of the 'Kyoto Six'.

To help achieve its objectives, Kyoto establishes three market-based 'flexible mechanisms'. These are: joint implementation of emissions reduction projects by two or more industrialised parties; the Clean Development Mechanism (CDM) under which emissions reduction projects can be implemented in developing state parties; and emissions trading. Guidelines for these mechanisms were fleshed out in the Marrakesh Accords agreed at the Conference of the Parties (COP) 7 in 2001, and formally adopted at the Meeting of the Parties (MOP) 1 in Montreal in 2005.[19]

6 The CDM and joint implementation

The most important flexibility mechanism to date is the CDM. Article 12 of the Kyoto Protocol establishes the CDM, the purpose of which is to help developing countries attain sustainable development, while also assisting Annex I parties in achieving compliance with their emissions limitation and reduction commitments. Hence Annex I states or private companies can acquire carbon credits known as Certified Emission Reductions from projects such as reforestation or afforestation in developing countries which can count towards their Kyoto obligations.

There has been a remarkable increase in CDM investment in recent years and there are now around 4000 CDM projects in operation or on the drawing board. Most CDM projects are in Asia and the Pacific region (65 per cent), with a sizeable share in Latin America and the Caribbean (32 per cent), and only a small fraction in Africa (2 per cent).[20] Although the premise of the CDM is that both developed and developing states will benefit, in reality major questions hang over the merits of the CDM. CDM projects must be entered into voluntarily by host states and they must achieve real reductions in emissions that are additional to those that would have been achieved but for the project. However, it has not been possible to establish with any certainty whether projects have achieved 'additionality'. More importantly, the CDM does not result in an aggregate reduction in emissions; it simply allows developed states to offset increases in their emissions by purchasing credits from developing states.

Joint implementation is similar to CDM. It is established by Article 6 of the Kyoto Protocol and allows Annex I countries to earn carbon credits known as Emission Reduction Units through investments in emission abatement projects in other Annex I countries. This will be desirable where the cost of abatement is much lower in one Annex I country compared to another, as is the case in 'economies in transition' such as states of the former Soviet Union.

7 Emissions trading

Emissions trading under the Kyoto Protocol, established by Article 17, is essentially a 'cap and trade' system similar to the Clean Energy Future emissions trading scheme to be implemented (discussed in the next chapter), except that it applies on a global scale and the participants in the scheme are nations rather than enterprises. Under the Kyoto emissions trading system, Annex I parties can meet their emission limitation and reduction targets by purchasing surplus emission units from other parties. The Kyoto Protocol does not set any numerical limit to the volume of these units (known as Assigned Amount Units) than may be purchased, but it does state that emissions trading (and the other flexibility mechanisms) are to be supplementary to, and not a substitute for, domestic action to reduce emissions.

The advantages and disadvantages of emissions trading schemes are discussed more fully in the next chapter, in the context of the options available to Australia. In broad terms, the benefits of global emissions trading are the same as they are for domestic trading schemes. Emissions trading is a market-based approach to mitigating climate change. It serves to place a price on carbon, providing a clear incentive for the market to turn to less carbon-intensive technologies. Emissions trading is challenging to administer, given the need to allocate permits and keep a track of their trade on the market, and therefore involves higher transaction costs than a carbon tax. There is also a range of possible emissions trading models, and whether or not it will be successful depends on the particular way it is designed.

Trading in emissions rights and in offsets under the Kyoto Protocol is in many respects in its infancy, but if there is an ambitious global deal to cut emissions, and targets set for individual countries above and beyond the modest limitation and reduction targets in the first Kyoto commitment period, then international emissions trading holds great promise in reducing global emissions. As Ross Garnaut has explained, it would mean that the lowest cost abatement opportunities are taken up, provide incentives for developing countries to take up emissions reduction targets, and set stable and consistent carbon prices worldwide.[21]

8 Kyoto compliance procedure

One of the innovative features of the Kyoto Protocol is its compliance procedure, which is designed to ensure that states meet the rules and commitments laid down by the treaty, and supply reliable data on their emissions. If a party fails to meet its obligations then the compliance system can be activated, which can lead to an enforcement branch taking action, such as suspending the entitlement of a party to participate in the flexibility mechanisms. If a party fails to meet an emissions limitation or reduction target, the enforcement branch may impose a penalty which involves increasing the party's emission limitation or reduction target by 30 per cent for the second commitment period.

The Kyoto Protocol's compliance procedure is one of a number of similar procedures established for environmental regimes in recent years, the first of which was the Montreal Protocol. They differ from traditional mechanisms for achieving compliance in international law, such as international litigation through courts such as the International Court of Justice, in that they are designed to be more proactive and facilitative, to give states a helping hand in meeting their obligations instead of making black-and-white findings that a rule has been broken.

To date there have been several compliance cases that have been taken to the enforcement branch. The first, in 2006, involved Greece, which did not have an adequate national system for estimation of anthropogenic emissions by sources and removals by sinks of greenhouse gases as required by Article 5 of the Kyoto Protocol. A case against Canada in 2008 concerned the adequacy of its 'national registry', the computerised system to track holdings of greenhouse gas credits required by Article 7. A third case, in 2009, involved Croatia, and was concerned with technical issues connected with the calculation of assigned emissions amounts. A more recent case involved Bulgaria, which was suspended from Kyoto Protocol emissions trading in mid-2010 because it did not have appropriate national systems in place. This had the effect of preventing Bulgarian companies subject to the European Union Emissions Trading Scheme from obtaining revenue from transferring European Union Emissions Allowances. Following a review of remedial measures in February 2011, Bulgaria regained its accreditation to participate in Kyoto carbon markets.

9 Post-Kyoto developments

While undoubtedly a major achievement in global environmental governance, in truth the Kyoto Protocol was a very modest first step towards combating climate change. The Kyoto Protocol's requirement that industrialised countries reduce or limit their emissions by around 5 per cent from those of 1990 translates into no more than a 1 per cent global reduction in emissions. This is a trifling abatement. However it must be remembered that this is only the initial target, for the first commitment period of 2008–12. When

Kyoto was negotiated it was anticipated that the first commitment period would be followed by second and subsequent periods that would ratchet up the emissions cuts to match what the science says needs to be done.

Since the Kyoto Protocol was concluded there has been a hiatus in the development of international climate change law. This continues because the Copenhagen Accord agreed in December 2009, and discussed below, is a non-binding text and the Cancún Agreements concluded in December 2010 fall short of a revised or new Kyoto Protocol. Nonetheless, there have been several interim developments of note, including the Asia Pacific Partnership on Clean Development and Climate.

10 The Asia Pacific Partnership on Clean Development and Climate (APP)

The APP was agreed in Sydney in 2006, and was initially an attempt to sideline the UNFCCC and Kyoto by two of the APP's main proponents – Australia and the United States. Both states at the time rejected the Kyoto Protocol (the United States still does).

Although presented as an agreement that is complementary with Kyoto, this non-binding partnership is in substance foreign to the principles that animate Kyoto and the UNFCCC. The APP comprises the United States, China, India, Japan, Korea and Australia, which together produce half of the world's global emissions. It embraces the tropes of voluntarism and technologism that have dominated the Australian government's domestic response to climate change (see further Chapter 4).

It sets no targets, timeframes or benchmarks for emissions reductions, but places its faith in promoting technological innovation through voluntary programs and government subsidies. There are a number of good reasons to think that the APP breaches the UNFCCC, not least because it promotes an increase in emissions so long as there is a decline in greenhouse gas intensity (that is, the level of emissions as a proportion of GDP), whereas the UNFCCC requires total emissions reductions. The APP also appears ineffective. According to modelling by the Australian Bureau of

Agricultural and Resource Economics it is unlikely to deliver any substantial emissions reductions.[22]

The APP survived a change of government in Australia in 2007 and continued to be supported by the new governments of Prime Ministers Rudd and Gillard. However, it is no longer presented as the 'main game' for climate negotiations – in this respect the Australian government has returned to multilateral institutions – accepting that the only effective arena for a global deal is under the UNFCCC framework.[23]

The non-binding and non-Kyoto compliant APP is to be distinguished from regional schemes such as the European Union Emissions Trading System which can, and do, make a contribution to addressing climate change consistent with the UNFCCC framework. The European Union system has been designed to link into Kyoto Protocol emissions trading. It provides the main regulatory scheme under which European nations have successfully reduced their emissions in line with the Kyoto target of an 8 per cent reduction on 1990 levels (which applies collectively for the 15 European states that were members of the European Union at the time the Kyoto Protocol was adopted in 1997).

11 The Copenhagen Accord

The choice facing the international community is whether to rollover Kyoto and adopt a follow-up framework that sets new emissions targets, that includes a wider range of emitters with greater stringency, or whether to discard Kyoto and adopt a new regime altogether. Given the international commitment to Kyoto and its long and difficult gestation, it would be naïve to think that a new agreement can easily be concluded. As Clive Hamilton has noted, Kyoto represented 'the most complex and ambitious international treaty process ever attempted'.[24] As such, consensus on its basic foundations, including the heavy lifting to be taken by developed countries, cannot readily be pushed aside.

At the Copenhagen climate conference in December 2009 a number of developed countries, including Australia, advocated the disbanding of Kyoto and its replacement with a new treaty imposing binding emission reduction cuts on both developed and developing countries. It was hoped that such an agreement would garner the support of the biggest emitters - China and the United States - without whose cooperation no climate deal is possible. However, the Group of 77 developing countries and China refused to countenance this.

The Copenhagen conference came at the end of two years of negotiations that began in 2007 with the conclusion of the Bali Action Plan. At previous climate conferences the main division has been between developed nations and developing nations - between the North and the South. In Copenhagen there was a new fracture, with a small group of developing and least developing states including Tuvalu standing up to China and India. This division reflected that the larger, wealthier developing countries were keen to preserve their ability to emit carbon during their economic development, whereas the smaller, least developed countries largely did not possess or rely on carbon intensive industries and were likely to be the most vulnerable to – and least able to cope with – the adverse impacts of climate change.

The divisions proved insurmountable, and Copenhagen led to no legally binding outcome whatsoever. Instead what was agreed was merely a 12-paragraph political 'accord' - the Copenhagen Accord - negotiated directly between President Barack Obama, Chinese Premier Wen Jiabao, Indian Prime Minister Manmohan Singh, Brazilian President Luiz Inacio Lula and South African President Jacob Zuma. The Accord was not formally adopted by the Conference of the Parties (COP) to the UNFCCC, or the Meeting of the Parties (MOP) to the Kyoto Protocol. Instead the COP and the MOP adopted decisions that they 'take note' of the accord, leaving governments then to sign on to the document if they wished to do so.

The device whereby the COP/MOP 'took note' of the Accord means that it does not even have the status of United Nations conference declarations such as the 1972 Stockholm Declaration and 1992 Rio Declaration that were adopted by consensus. These declarations are often described as 'soft law', which means that they set down general normative standards that can be used to guide state behaviour, but they are not legally binding in the same way as a treaty is. The Copenhagen Accord is even softer law. Rather than being drafted by the COP and MOP as a whole it was composed by a small number of states, in a process that was criticised by many governments as being non-transparent.

The Copenhagen conference attracted enormous global media and public attention and was seen by many as the 'make or break' event for responding to climate change before it is too late. In addition to 10,500 official government delegates, there were around 1300 registered non-governmental organisations (NGOs), 13,500 observers, and more than 3000 media personnel from around the world. The high hopes generated in the lead up to Copenhagen soon turned to widespread disappointment, and even anger, as the conference failed to deliver on many people's aspirations. Given the complexity of the problem, the large number of delegates and the short time available at Copenhagen, and the long periods needed to build global political consensus, expectations for the conference were perhaps unrealistically high, if understandable.

Was Copenhagen the failure it is popularly portrayed as? Although there was a slow start in UNFCCC parties associating themselves with the Copenhagen Accord, it did ultimately attract a good deal of support. By February 2011, 138 states, including the European Union, had engaged with the Accord, representing 86.76 per cent of global emissions.[25] Hence it appears that, although the majority of nations support the Accord, it will not have the same universal support as the UNFCCC and Kyoto. Indeed seven countries to date have specifically stated that they do not wish to engage with the Accord,[26] reflecting the dissatisfaction among some developing states about the non-transparent manner in which the Copenhagen Accord was drafted.

There are both positive and negative features of the Copenhagen Accord. On the positive side of the ledger, it is significant that the agreement is more inclusive than the Kyoto Protocol, in that it brings together developed and developing states for the first time in terms of mitigation commitments. It was also significant that the Copenhagen Accord puts a dollar figure on the funds to be made available by developed states to meet adaptation and mitigation needs in developing countries, although this is only a fraction of what is required in the long term. In terms of its substantive provisions, the Copenhagen Accord addresses the following issues, which are also taken up in the subsequent Cancún Agreements.

(a) Mitigation

The Accord notes the 'scientific view that the increase in global temperature should be below 2 degrees Celsius'. Although the Accord does not outline what global emission cuts are required to meet this goal, it does state that 'deep cuts are required according to science' and cuts of 50–60 per cent are essentially implied by the 2°C limit. In addition the last paragraph recognises that even this limit is controversial, and calls for the assessment of the implementation of the Accord by 2015, including a consideration of strengthening the long-term goal to limit temperature rises to 1.5°C.

The Accord does not set out what individual countries must do in terms of abating emissions. However, it does provide that Annex I (developed) countries must commit individually or jointly to economy-wide emission targets for 2020 to be submitted to the UNFCCC Secretariat by 31 January 2010 for compilation in an appendix. Non-Annex I (developing) countries are required to take 'mitigation actions' also to be added to the appendix. Such measures could include reducing emissions intensity per unit of gross domestic product, rather than aggregate emissions cuts. Although the 31 January deadline was missed by many nations, some major developed country emitters have indicated what their emissions cuts will be. The United States, for example, pledged to cut its greenhouse gas emissions by 17 per cent by 2020 from 2005 levels, although it made its commitment contingent on passing legislation at home. The European Union's pledge was a 20–30 per cent reduction by 2020 from a 1990 baseline. The big developing economies, China and India, made commitments to reduce carbon intensity, rather than absolute emissions reductions. For China this was 40–45 per cent and India 20–25 per cent.

Emissions targets set by Annex I countries are to be subject to rigorous, robust and transparent measurement, reporting and verification. Thanks in part to the blocking by China, mitigation actions by developing countries are not subject to the same scrutiny – actions are to be subject to domestic measurement, reporting and verification, with the information to be reported to the UNFCCC

Secretariat and then subject to 'international consultation and analysis under clearly defined guidelines that will ensure that national sovereignty is respected'.

(b) Adaptation

The Accord included very little coverage of adaptation issues. Developed countries are to 'provide adequate, predictable and sustainable financial resources, technology and capacity-building to support the implementation of adaption action in developing countries'. As noted below, however, firmer agreement was reached on the global financing of climate change responses, including adaptation.

(c) Finance

Many developing countries are unable to take concrete steps to cut their emissions or build their adaptive capacity to climate change because they lack access to the necessary financial capacity. The Copenhagen Accord sought to fill the financing gap. 'Scaled up, new and additional, predictable and adequate funding' must be provided to developing countries to support mitigation efforts, adaptation, technology development and capacity building. A figure of US$30 billion is put on this – for the period 2010–12 this is the 'collective commitment' by developed countries. In addition developed countries committed jointly to mobilising US$100 billion a year by 2020 from both public and private sources. A significant proportion of this should come from a new fund to be known as the Copenhagen Green Climate Fund. These funding commitments were reaffirmed at the Cancún climate change talks in December 2010 and, while significant funds have been allocated by governments to the US$30 billon 'fast track' financing fund, questions hang over whether the funding will be sufficient to cover the needs of developing countries.

(d) Forestry

Emissions from deforestation account for up to 17 per cent of global greenhouse gas emissions from human activities annually[27] and the Copenhagen Accord seeks to address this problem for the first time.

To this end, the Accord declared the 'immediate establishment of a mechanism including REDD-plus, to enable the mobilization of financial resources from developed countries'. REDD refers to Reduced Emissions from Deforestation and Degradation and involves the use of public funding or market-based mechanisms to channel financing to developing states, such as Indonesia, to provide an incentive to reduce rates of land clearing.[28]

(e) Technology mechanism

Developing countries will only be able to reduce emissions if they are able to gain access to technological developments in industrialised nations that provide renewable sources of power and reduce emissions from industry. The Copenhagen Accord states that the parties decided to establish a Technology Mechanism to accelerate technology development to support action on both adaptation and mitigation. Ordinarily, the disparities in human and financial resources available for research and development of new technologies, and intellectual property law protections, place developing countries at a serious disadvantage compared with developed countries.

12 Post-Copenhagen developments

The Copenhagen Accord was important in so far as it is the first agreement to specify emissions reduction targets and actions by both developed and developing countries, and includes a target to limit warming to less than 2°C. However, it is not a strong, legally binding agreement that advances the principles of the UNFCCC.

It was hoped that the Copenhagen Accord would be the basis of a legally binding text agreed at the subsequent COP, COP 16, held at Cancún in Mexico in December 2010. The Cancún talks were certainly less fractious and more productive than those at Copenhagen, and signalled a reaffirmation of the United Nations as the multilateral institution through which a climate deal has to be brokered. The conference formally adopted a number of decisions known as the Cancún Agreements. These are considered by almost all UNFCCC states parties (with the sole exception of Bolivia) as a legitimate expression of the will of the international community, certainly more so than the Copenhagen Accord, even though the Cancún Agreements largely repeat and expand upon the core elements of the Copenhagen Accord.

At Cancún, as at previous climate conferences, the discussion proceeded along two tracks – the UNFCCC track which included the United States, and the Kyoto track which did not. Under the UNFCCC negotiation track, the Cancún Agreements included a decision to hold temperature increases to below 2°C, with this goal to be kept under periodic review with the potential to lower the increase to 1.5°C. The parties also agreed to improve the reporting, monitoring and verification of the mitigation contributions made by developed and developing countries. The idea of a REDD scheme was also acknowledged and supported.

Under the Kyoto track, there was, regrettably, no agreement on a second commitment period for the Protocol, which expires at the end of 2012. If a new agreement is not reached before then, the Kyoto Protocol will effectively cease to function, and the CDM and other emissions trading systems will no longer operate. The Cancún Agreements nonetheless set the scene for agreement at a conference in Durban in December 2011 on a second commitment period, which would include amendments to Annex B of the Kyoto Protocol. This is because they specifically acknowledge the importance of adopting new individual targets in the context in which developed countries will have to reduce their emissions by between 25 and 40 per cent on 1990 levels by 2020. It also stated that emissions trading under the Kyoto Protocol should continue.

At the Durban COP in December 2011 the 'Durban Platform for Enhanced Action' was agreed. This was not the reinvigoration of the Kyoto Protocol that many had hoped it would be. There was no agreement on emissions reductions for a second Kyoto commitment period. Nonetheless the Durban Platform is important because it keeps the Kyoto Protocol alive (if only on life support), calls for developed *and* developing countries to reduce their emissions, and sets a timeline for a new round of negotiations to be concluded by 2015 on a 'protocol, another legal instrument or an agreed outcome with legal force' that would have effect from 2020. Whether this will be achieved remains to be seen, and there are reasons to be doubtful, particularly given Canada's remarkable decision to denounce and withdraw from the Kyoto Protocol just a day after the Durban climate talks concluded. Legitimate questions may be asked whether the UNFCCC process will ever bear fruit. Although there are examples of successful multilateral environmental agreements, including the Montreal Protocol, and in other areas such as the law of the sea where the United Nations Convention on the Law of the Sea (UNCLOS) is rightly credited as having significantly reduced marine pollution and ocean dumping, climate change is something of a special case. As such it demands the examination of cooperation strategies alongside the UNFCCC that are complementary to it, but which do not have the same inherent problems.

D Complementary legal regimes

International environmental law has developed in a mostly sectoral and ad hoc manner. Regimes have been devised to address specific global or regional environmental problems, such as particular sources and types of transboundary pollution, rather than to promote transboundary environmental governance in a holistic and integrated manner. However, there is a significant degree of crossover or overlap between various environmental regimes. This means that a range of environmental treaties have potential application to climate change issues. A full review of these is beyond the scope of this chapter, but several deserve particular mention.

1 Marine pollution regimes

Greenhouse gases causing climate change have an impact on the marine environment, in terms of rising temperatures, rising sea-levels and also in driving ocean acidification. Ocean acidification is caused when CO_2 dissolves in seawater forming acid, which removes carbonate ions from the water which are utilised by many marine organisms to form calcite shells, such as planktons and molluscs.[29] Ocean acidification could be made worse by geo-engineering techniques such as the fertilisation of the oceans by iron compounds which would result in increased oceanic uptake of carbon dioxide.

The international regulatory framework for marine environmental protection is dominated by broad principles of environmental protection, such as those contained in Part XII of the UNCLOS, combined with denser patches of regulation where states have negotiated at global and regional levels on specific issues.

The UNCLOS imposes a general duty on states to protect and preserve the marine environment, and requires parties to take those measures necessary to prevent, reduce and control pollution of the marine environment from any source, and to adopt laws and regulations to prevent, reduce and control pollution of the marine environment from or through the atmosphere. The UNCLOS is supplemented by more detailed regimes, including those regulating dumping at sea and land and atmospheric source marine pollution, both of which are applicable to climate change to some extent.

The dumping regime (found in the 1972 London Dumping Convention and the 1996 London Protocol) applies to the sequestration of CO_2 in the oceans, which can be used to mitigate atmospheric CO_2 emissions. The dumping regime has developed in such a way as to permit sub-seabed sequestration of CO_2 but to prohibit water-column disposal of CO_2. In November 2006, amendments to the 1996 London Protocol were adopted permitting the storage of CO_2 under the seabed.[30] These allow for regulation of sub-seabed sequestration of CO_2, with CO_2 streams from carbon capture processes added to Annex I as a waste or other matter that may be considered for dumping. The resolution adopted by the parties in October 2008 that agreed that the scope of the London Convention and London Protocol includes ocean fertilisation activities and that, given the present state of knowledge, ocean fertilisation activities other than legitimate scientific research should not be allowed.[31]

2 Convention on Biological Diversity (CBD)

As noted in Chapter 1, climate change is having very substantial impacts on the biological diversity of both plants and animals, including by accelerating the rate of species extinction. The three main goals of the 1992 CBD, to which Australia is party, are the conservation of biodiversity, the sustainable use of the components of biodiversity, and the fair and equitable sharing of benefits arising from the commercial and other utilisation of genetic resources.

Several provisions of the CBD are relevant to climate change. Article 6 requires parties to develop national biodiversity strategies and action plans, and to integrate these into broader national plans. Article 7 requires parties to identify and monitor processes and categories of activities that have, or are likely to have, adverse impacts on the conservation and sustainable use of biological diversity and monitor their effects.

At the ninth meeting of the COP to the CBD, held in Bonn in 2008, a number of decisions were adopted which included reference to climate change issues. In the key decision, the COP called for coordinated action to address the threat that climate change poses to biodiversity.[32] Specific attention was directed at the question of ocean fertilisation, with the COP urging the parties to act in accordance with the decision of the 1972 London Convention and 1996 London Protocol, and requested parties, in accordance with the precautionary approach, to ensure that ocean fertilisation activities do not take place until there is an adequate scientific basis on which to justify such activities, including assessing associated risks.[33]

3 World Heritage Convention

The 1972 Convention Concerning the Protection of the World Cultural and Natural Heritage ('World Heritage Convention') is one of the earliest and most important environmental treaties. Australia is a party to the World Heritage Convention and, on the basis of the convention, the Commonwealth government has acted to conserve various natural and cultural heritage sites, including the South West Tasmanian wilderness which was protected from a dam project proposed by the Tasmanian government in the 1980s.[34]

One of the best known World Heritage sites is the Great Barrier Reef which is under direct threat from climate change, as rising water temperatures cause a phenomenon known as coral bleaching when the coral eject the coloured algae living within the coral tissue.[35] The Great Barrier Reef is the world's largest World Heritage property and, under the World Heritage Convention, Australia has to meet demanding obligations to ensure that the site is conserved, protected and transmitted to future generations.[36] The World Heritage Convention imposes an obligation of result and it is arguable that all possible steps to preserve natural heritage should be taken, including not only steps to improve the resilience of properties (in the case of the Great Barrier Reef by reducing other stressors on the reef such as pollution) but also to reduce its own greenhouse gas emissions and to press for cooperation on global reductions – the only sure way of protecting the reef in the long term.

4 Trade law

Governments may begin to lose patience with the slow pace of international negotiations on emissions reductions and seek to forge ahead with unilateral climate policies in an effort to kick start greater international action. Administrations in both the European Union and the United States have indicated that they will consider imposing trade barriers on carbon-intensive imports, unless the states where those products are produced to take concrete steps to curb their emissions.

The problem with such unilateral measures is that they may conflict with the requirements of international trade law. International trade law has developed since the 1940s to promote greater levels of global trade, free of protectionist restrictions. It became more institutionalised in 1995,[37] with the establishment of the World Trade Organization (WTO). The WTO provides a powerful institutional basis for implementing the international trade rules that had developed out of the General Agreement on Tariffs and Trade (GATT)[38] after the Second World War.[39]

Integral to the WTO is a compulsory and binding system of dispute settlement established by the Understanding on Rules and Procedures Governing the Settlement of Disputes, which affords broad rights of standing to WTO members to challenge the legality of actions by other members.[40] Under this procedure members may initiate consultations and, should they fail, request the establishment of a Panel.[41] Appeals against Panel reports may thereafter be brought before a standing Appellate Body.[42]

Since it commenced operation in 1995, over 356 complaints have been notified to the WTO, with almost 200 Panel and Appellate Body reports adopted.[43] The system has encountered environmental issues, with a number of disputes involving essentially environmental questions being brought before GATT panels and the WTO. This has in turn implicated environmental rules and standards.

The key organising principles of the WTO system are the most favoured nation principle[44] and the national treatment principle.[45] The most favoured nation principle requires equality of treatment for like products and, therefore, members of the WTO may not discriminate between like products of other members. The national treatment principle stipulates that domestic and imported products should be treated equally in terms of the application of internal regulations and policies. Together these disciplines limit the circumstances in which states may impose trade restrictive measures. Various national climate change policies may conflict with WTO rules applying to tariffs (border measures), the imposition of quotas, the granting of subsidies and labelling of products. However, they are subject to Article XX of the GATT, which sets out certain limited exceptions for health and environmental measures:

> Subject to the requirement that such measures are not applied in a manner which would constitute a means of arbitrary or unjustifiable discrimination between countries where the same conditions prevail, or a disguised restriction on international trade, nothing in this Agreement shall be construed to prevent the adoption or enforcement by any contracting party of measures: …
>
> (b) necessary to protect human, animal or plant life or health; …
>
> (g) relating to the conservation of exhaustible natural resources if such measures are made effective in conjunction with restrictions on domestic production or consumption.[46]

The case law to date in the WTO has been promising in so far as it has recognised the capacity of governments to introduce measures having a restrictive effect on trade so long as these are justified on the basis of the protection of the environment.[47] In this regard, the WTO Appellate Body has been responsible for a substantial 'greening' of WTO law, certainly in comparison with decisions of ad hoc panels established under the previous GATT regime. There is no reason to think that disputes concerning climate change measures would be approached any differently.[48] However, the outcome of a case will depend largely on the nature of the national measure that is implemented.

A measure that is likely to be introduced will be carbon taxes combined with border tax arrangements to ensure that domestic producers are not placed at a competitive disadvantage. A border tax arrangement would impose on imports a carbon tax equal to that which would have been applied had the import been produced domestically. Theoretically, it is possible to tailor these arrangements in such a way so as not to infringe the WTO principles of national treatment and most favoured nation status. However, there remains considerable room for debate over the legality of particular measures on the basis of matters such as the method of calculating the greenhouse gas emissions resulting from the manufacture of a given product.[49]

To avoid uncertainty and friction Ross Garnaut has recommended that there should be a new WTO code on the subject.[50] Garnaut notes that it would be difficult to achieve this as support from WTO members would need to be unanimous. However, 'given combined [European Union-United States] leadership, the credible threat of unilateral responses if no agreement were reached, the other incentives for cooperation on climate change, and the strong United Nations role ... an agreement may be possible'.[51]

E International climate change litigation

Finally, one rather ad hoc legal approach to dealing with climate change has been through litigation in domestic[52] and international courts and complaints bodies. It is possible to identify three types of international proceedings that may involve Australia. The first are what might be termed 'progressive' proceedings designed to bring positive outcomes for global climate policy by seeking to enforce emissions cuts in an international court or tribunal. Australia could conceivably be an applicant or a respondent in such proceedings. Australia has accepted the jurisdiction of a number of international courts and tribunals that could hear a case involving liability for climate change, including the International Court of Justice and the International Tribunal for the Law of the Sea.

A second category of international litigation can be termed 'regressive', in the sense that it may be brought to prevent Australia or other states from implementing policies to reduce emissions if these had the effect of interfering with other norms such as those relating to trade liberalisation. A third category of international proceedings more likely to be pursued than the former two are 'administrative' proceedings such as those pursued under the compliance regime established by the Kyoto Protocol to ensure that Australia adheres to its obligations.[53]

In relation to the first category of 'progressive' proceedings, there are several norms of both customary and conventional character that could be used by states impacted by climate change as a basis for proceedings against states that have failed to take steps to limit greenhouse gas emissions. The Pacific microstate Tuvalu, for instance, which is directly impacted by rising sea-levels, declared on signing and ratifying the UNFCCC and Kyoto Protocol that joining the climate regime 'shall in no way constitute a renunciation of any rights under international law concerning state responsibility for the adverse effects of climate change'. As with domestic tort law proceedings, it is unlikely that international proceedings based on state responsibility for transboundary damage are likely to be successful, primarily for reasons surrounding the difficulty determining causation for specific climate change impacts.

Potentially more promising 'progressive' litigation is that which can make specific reference to treaty provisions that have a bearing on climate change policy. An example is the World Heritage Convention,[54] which imposes obligations on states parties such as Australia to protect and transmit to future generations outstanding world cultural and natural heritage.[55] The two World Heritage sites in Australia most vulnerable to climate change are the Great Barrier Reef (which has already experienced widespread coral bleaching as a result of elevated water temperatures) and Kakadu National Park (which is threatened by rising sea-levels). Another example of 'progressive' proceedings are those that could be pursued in human rights complaints bodies such as the United Nations Human Rights Committee, which Australia has recognised as having competence to examine individual human rights communications from those whose rights are violated by Australia.[56]

'Regressive' proceedings through the WTO dispute settlement system appear increasingly likely if states seek to take unilateral measures to impose trade restrictions on the import of carbon-intensive products. Although the WTO has developed an important body of jurisprudence which has accepted the legitimacy of certain unilateral trade measures designed to protect exhaustible natural resources,[57] how it would address a measure such as a border tax arrangement implemented as part of a national emissions trading scheme remains highly uncertain. The attitude of the Australian government to such measures is also unclear, as it has generally been a strong and active supporter of the global trade liberalisation agenda, and has participated as a third party in proceedings in the WTO supporting challenges to several measures such as the European Union's moratorium on the import of genetically modified agricultural products.[58]

Turning finally to 'administrative' proceedings within the climate change regime itself, Australia could be the target of proceedings if it fails to meet its procedural or substantive obligations under the Kyoto Protocol or its successor. The Kyoto Protocol compliance procedure has begun to be utilised to deal with a number of what are termed by the regime to be 'questions of implementation', with alleged or actual non-compliance by several states having been considered by the Enforcement Branch of the compliance system.

F Conclusion

The international law of climate change is now almost two decades old. The UNFCCC was a significant international environmental agreement, reached just two years after the first assessment report of the IPCC in 1990 confirmed the seriousness of human-induced climate change. The UNFCCC sets out the basic principles that continue to animate international climate law. The Kyoto Protocol was also a significant accomplishment, requiring industrialised states for the first time to meet legally binding emission cuts. However, the achievement at Kyoto was significantly undermined by the United States and by Australia, which slowed down international progress towards a more effective regime.

Since the election of President Barack Obama in 2008 there have been significant efforts made to play catch up in international climate law, but it is unrealistic to think that the intransigence can be overcome quickly. In addition, there is a new dynamic in the geopolitics of climate change, not apparent in the 1990s, which is the rise of the rapidly industrialising states China and India that are now playing the role of intransigents. China, in particular, holds the key to an effective post-Kyoto deal. Unless and until it can be persuaded or induced to enter into an agreement it will not be possible to move forward and make the emission reductions necessary to avoid dangerous change.

Given the urgency of reducing emissions the challenge of reaching an ideal international response should not be used as an excuse for inaction. Indeed it is only through states such as Australia, which has among the highest per capita emissions in the world, taking meaningful steps to reduce their own emissions that it will be possible to demonstrate that the developed world is acting to redress its historical responsibility for the problem, leaving the way open for China in particular to address its contemporaneous, and future, responsibility.

Australia and other like-minded nations should also consider taking action on a regional, bilateral and even unilateral basis in order to kick start progress on a truly international response to climate change. Action such as imposing trade barriers on products from nations failing to live up to the challenge of climate change may be controversial, and would almost certainly be the subject of litigation – however, all options must be kept on the table. The next chapter explores the policy options available to Australia to show leadership at home in addressing climate change.

Notes

1 Garnaut, R, 2008, *The Garnaut Climate Change Review: Final Report*, Cambridge University Press, Cambridge, p 173.

2 Ibid, p 89.

3 The Alliance of Small Island States is a grouping of 39 island states in the Pacific, the Indian Ocean and the Caribbean.

4 See McNeil, B, 2009, *The Clean Industrial Revolution: Growing Australian Prosperity in a Greenhouse Age*, Allen & Unwin, Sydney.

5 Baumert, KA, et al, 2005, *Navigating the Numbers: Greenhouse Gas Data and International Climate Policy*, World Resources Institute, Washington DC, 32.

6 Meyer, A, 2000, 'Contraction and Convergence' 157 *Engineering Sustainability* 189.

7 Garnaut, above n 1, pp 206–07.

8 Ibid, p 204.

9 *Trail Smelter Case (Canada/United States of America)* (1938 & 1941) 3 RIAA 1911. For discussion, see Stephens, T, 2009, *International Courts and Environmental Protection*, Cambridge University Press, Cambridge, pp 123–36.

10 Ibid, p 1965.

11 Declaration of the UN Conference on the Human Environment, UN Doc A/CONF.48/14/Rev.1 (1973).

12 UN Declaration on Environment and Development, UN Doc A/CONF.151/5/Rev.1 (1992).

13 See Verheyen, R, 2005, *Climate Change Damage and International Law: Prevention Duties and State Responsibility*, Martinus Nijhoff Publishers, Leiden.

14 UN, 2009, *Millennium Development Goals Report 2009*, United Nations, New York, p 41.

15 UN Environment Programme Ozone Secretariat, 2009, *The Montreal Protocol on Substances that Deplete the Ozone Layer: Achievements in Stratospheric Ozone Protection: Progress Report 1987–2007*, United Nations, New York.

16 For a first-hand account, see Sands, P, 2005, *Lawless World: Making and Breaking Global Rules: From FDR's Atlantic Charter to George W. Bush's Illegal War*, Penguin Group, New York, pp 76–94.

17 Article 4(2)(a) and (b).

18 Article 3(1).

19 For a comprehensive overview of these, see Brohé, A, Eyre, N and Howarth, N, 2009, *Carbon Markets: An International Business Guide*, Earthscan Publications Ltd, London/Sterling.

20 Garnaut, above n 1, p 182.

21 Ibid, p 228.

22 Fisher, B et al, 2006, *Technological Development and Economic Growth*, Australian Bureau of Agriculture and Resource Economics Research Report 06.1, Inaugural Ministerial Meeting of the Asia Pacific Partnership on Clean Development and Climate, Sydney, p 34. See also Hamilton, C, June 2006, *The Political Economy of Climate Change*, Milthorpe Lecture, Macquarie University, available at <http://www.tai.org.au>.

23 See the discussion in Lawrence, P, 2009, 'Australian Climate Policy and the Asia Pacific Partnership on Clean Development and Climate: From Howard to Rudd, Continuity or Change?' 9 *International Environmental Agreements* 281.

24 Hamilton, C, 2007, 'Building on Kyoto' 45 *New Left Review* 91 at 96.

25 See <http://www.usclimatenetwork.org/policy/copenhagen-accord-commitments>.

26 Bolivia, Cuba, Ecuador, Kuwait, Nauru, Tuvalu and Venezuela.

27 Garnaut, above n 1, p 235.

28 See Lyster, R, 2009, 'The New Frontier of Climate Law: Reducing Emissions from Deforestation and Degradation' 26 *Environmental and Planning Law Journal* 417.

29 See the discussion in Baird, R, Simons, M and Stephens, T, 2009, 'Ocean Acidification: A Litmus Test for International Law' 3 *Carbon and Climate Law Review* 459.

30 IMO Doc LC-LP.1/Circ.5 (27 November 2006).

31 IMO Res LC-LP.1 (31 October 2008).

32 COP IX/16 Biodiversity and Climate Change, UNEP/CBD/COP/DEC/IX/16 (9 October 2008).

33 COP IX/20 Marine and Coastal Biodiversity, UNEP/CBD/COP/DEC/IX/20 (9 October 2008).

34 Boer, B and Wiffen, G, 2006, *Heritage Law in Australia*, Oxford University Press, New York, pp 97–98.

35 Australian Government, Department of Sustainability, Environment, Water, Population and Communities, 2009, *Implications of Climate Change for Australia's World Heritage Properties: A Preliminary Assessment*, Australian National University, Canberra, p 53.

36 See Articles 4–7.

37 Marrakesh Agreement Establishing the World Trade Organization, 15 April 1994, 1867 UNTS 154, 33 ILM 1144 (1994) ('1994 Marrakesh Agreement').

38 1947 General Agreement on Tariffs and Trade. See now the 1994 Marrakesh Agreement, annex 1A: 1994 General Agreements on Tariffs and Trade, 15 April 1994, 1867 UNTS 187, 33 ILM 1153 (1994) ('GATT 1994').

39 See generally Goh, G and Witbreuk, T, 2001, 'An Introduction to the WTO Dispute Settlement System' 30 *University of Western Australia Law Review* 51 at 52.

40 1994 Marrekesh Agreement, annex 2 (Understanding on Rules and Procedures Governing the Settlement of Disputes).

41 1994 Marrakesh Agreement, Articles 4(7) and 5(4).

42 1994 Marrakesh Agreement, Article 17.

43 Leitner, K and Lester, S, 2007, 'WTO Dispute Settlement From 1995 to 2006 – A Statistical Analysis' 10 *Journal of International Economic Law* 165.

44 GATT 1994, Article I.

45 GATT 1994, Article III.

46 Given the relatively recent emergence of environmental concerns in international law it is unsurprising that the only reference of the GATT is to 'natural resources' rather than the environment more broadly: Harris, M, 2003, 'Beyond Doha: Clarifying the Role of the WTO in Determining Trade-Environment Disputes' 21 *Law in Context* 307 at 309.

47 See especially *United States – Import Prohibition of Certain Shrimp and Shrimp Products*, WTO Doc WT/DS58/AB/R. See the discussion in Stephens, above n 9, pp 321–42.

48 See generally Triggs, G, 2002, 'Kyoto Protocol: Compliance and the World Trade Organization' *Australian Mining and Petroleum Association Yearbook* 49.

49 See Genasci, M, 2007, 'Border Tax Adjustments and Emissions Trading: The Implications of International Trade Law for Policy Design' 1 *Carbon and Climate Law Review* 33.

50 Garnaut, above n 1, p 233.

51 Ibid, pp 233–34.

52 In relation to the litigation to date in Australian courts, see Preston, BJ, 'Climate Change Litigation' in Lyster, R (ed), 2010, *In the Wilds of Climate Law*, Australian Academic Press, Brisbane, p 208.

53 See further Stephens, T, 'International Courts and Climate Change: "Progression", "Regression" and "Administration"' in Lyster (ed), above n 52, p 53.

54 Opened for signature 16 November 1972, [1975] ATS 47 (entered into force 16 December 1975).

55 Articles 4 and 5.

56 See further Cordes-Holland, O, 2008, 'The Sinking of the Strait: The Implications of Climate Change for Torres Strait Islanders' human rights protected by the ICCPR' 9 *Melbourne Journal of International Law* 405.

57 See Stephens, above n 9, ch 10.

58 *European Communities – Measures Affecting the Approval and Marketing of Biotech Products*, Report of the Panel, WTO Doc WT/DS291/R, WT/DS292/R, WT/DS293/R (2006).

Chapter 4

Australia's Response to Climate Change

Key Issues

- Australia was a strong supporter of the United Nations Framework Convention on Climate Change (UNFCCC), though this instrument imposed few concrete obligations. The UNFCCC's main aim of reducing carbon emissions by 20 per cent from 1998 levels by 2005 was not enacted into Australian law or policy.
- The Howard Coalition government refused to sign the Kyoto Protocol until Australia was permitted to increase its emissions and granted special concessions relating to land use. Even after special concessions were made, Australia refused to ratify it. Ratification was the new Rudd Labor government's first official act after its election in 2007, followed by the creation of a mandatory, enforceable emissions reporting system for companies.
- At the 2009 Copenhagen conference, Australia advocated a new, legally binding treaty to reduce emissions that would replace the Kyoto Protocol and bring developed and developing states under the same regime. Australia nonetheless supported the non-binding Copenhagen Accord, especially that global temperature increases had to be limited to no more than 2°C and that both developed and developing countries should act.
- Australia's emissions reduction targets at Copenhagen were not ambitious and merely restated its pre-Copenhagen targets. Since the 2010 Cancún Agreements, Australia has adopted a cautious 'wait and see' approach committing to further reduce its emissions.
- Given its historical involvement in causing climate change, its high per capita emissions and its relative prosperity, Australia should act before developing countries in reducing emissions. Although the case for acting before other developed countries is weaker, many other developed countries have already taken strong action to reduce their emissions.

- Economic analyses indicate that Australia can easily afford to reduce its overall emissions in line with its international obligations and 'best practice' policy recommendations. Australia can do this by carrying out emission-reducing activities; mandating firms to carry out such activities; or imposing taxes and liabilities to encourage such actions.
- The most cost-effective methods are market-driven mechanisms (such as emissions trading) with few exemptions. Some concessions (such as limited free permits) may be necessary in the short term to compensate for loss of asset value and to sustain the global competitiveness of certain exposed industries.
- Between 1996 and mid-2007 the Howard government took little serious action to develop a comprehensive domestic policy on climate change. In the face of rising opposition in the lead-up to an election, in late 2007 the Howard government endorsed a proposal for an Australian 'cap and trade' emissions trading scheme by 2012.
- After the 2007 election, the Rudd government's pledge to establish an Emissions Trading Scheme (ETS) was fraught with difficulty. Its proposed Carbon Pollution Reduction Scheme (CPRS) made various concessions to industry groups, undermining its environmental and economic integrity. The revised CPRS Bill was defeated in the Senate in August 2009 and in April 2010 the government suspended the reintroduction of a CPRS until 2012.
- The failure of the CPRS partly explained Rudd's replacement as Prime Minister by Julia Gillard in June 2010. After the 2010 election, Gillard established a Multi-Party Climate Change Committee in September 2010 and later endorsed its strategy on carbon pricing.
- In July 2011 the Gillard government announced its 'Clean Energy Future' plan. Central to it was the introduction of a carbon price, with a fixed price mechanism (like a tax, with carbon priced at $23 per tonne) from July 2012, followed by a flexible market mechanism (an emissions trading scheme) from July 2015. Around 500 of Australia's 'biggest polluters' will need to buy a permit for each tonne of pollution emitted, as an incentive to reduce their pollution. Funds raised would compensate households for cost of living increases.

- The Clean Energy Future plan is a good start because it prices carbon pollution and will begin to reduce emissions, while mitigating social justice costs through compensation. From an environmental and economic stand-point, it is not optimal because it gives too many exemptions to too many industries or polluting activities. But it is still far more cost effective and likely to reduce emissions than the Opposition's 'direct action' plan.
- Options exist for significantly reducing emissions from each major sector. For electricity these include growth of both non-renewable energy sources (such as natural gas or nuclear power) and renewable ones (such as wind and solar). Transport emissions can be reduced through alternative energies (such as electricity or non-food bio-fuels), greater fuel efficiency and improvements in public transport and urban design. Better heating, cooling and lighting design can reduce the energy consumed by buildings. Emissions can be reduced from forestry and agriculture by avoiding deforestation, pursuing reafforestation and investing in new farming technologies. Changes to individual behaviour, including shopping or spending priorities, can also significantly reduce emissions.

A Introduction

As in the international arena and in many other countries, policy debates about climate change have been controversial in Australia. Initially, Australia supported international efforts to combat climate change by advocating for the adoption of the United Nations Framework Convention on Climate Change (UNFCCC) between 1988 and 1992, although domestically Australia's response was less productive in practice. A change of government in 1996 saw Australia engage in negotiations for the Kyoto Protocol, yet Australia drove a hard bargain (to extract special concessions for itself) and then refused to become a party to the treaty.

Australia's participation in the Kyoto Protocol eventually came about with a change of government in 2007 and Australia also supported negotiations for a global climate deal at Copenhagen and beyond. From around 2007 onwards, debates about Australia's domestic policy responses intensified, as the new government struggled to design a mechanism for reducing Australia's carbon emissions. Those domestic difficulties, related to problems at the international level, resulted in the government postponing its plans to deal with climate change. Public dissatisfaction with that deferral was one factor leading to a change of Prime Minister in 2010. The new Prime Minister sought to reinvigorate climate change policy in 2011, but immediately faced significant political opposition, ensuring the issue remained deeply politicised.

This chapter first considers Australia's participation in international negotiations about climate change and the various positions successive Australian governments have taken on key issues. It then reviews the domestic policy options available to Australia in confronting climate change, including a carbon tax and different types of emissions trading schemes (such as 'cap and trade' or 'baseline and credit'). The discussion considers the extent to which the different options could help to meet Australia's international obligations and support future international cooperation. It also examines the advantages and disadvantages of different mechanisms, including their environmental effectiveness (that is, how far they reduce emissions) and economic efficiency (how cheaply they reduce emissions).

B Australia, the UNFCCC and the Kyoto Protocol

Australia was one of the first states to join the UNFCCC and is an Annex I party to that treaty (the previous chapter explains the UNFCCC and the Kyoto Protocol which followed it). The Australian government, led by then Prime Minister Bob Hawke, played a supportive role in the negotiation of the UNFCCC during the 1980s, including through the office of the Ambassador for the Environment which was created in 1988. That post was first held by Sir Ninian Stephen, a former Justice of the High Court and Governor-General.

The UNFCCC imposes few concrete obligations on state parties, other than to collect and report emissions data. The Hawke government adopted as an Interim Planning Target the emissions reduction target recommended by an international conference in 1988. This required the reduction of carbon dioxide emissions by 20 per cent from 1998 levels by 2005 and, ultimately, by 50 per cent.[1] However, this objective was not entrenched in Australian policy or law, and was never met in any event (Australia's emissions increased by 4.2 per cent from 1990 to 2006).[2]

In 1996, the Howard government was elected and adopted a far less supportive attitude to climate negotiations than the predecessor Hawke/Keating governments. The new government participated in the international climate negotiations at Kyoto, led by the Minister for the Environment, Senator Robert Hill. Senator Hill drove a tough bargain at the Kyoto conference, refusing during the last, all-night conference to sign the agreement being negotiated unless two key concessions were granted to Australia.[3]

The first of these concessions was that Australia be permitted to *increase* its emissions by 8 per cent over 1990 levels, during the first commitment period of 2008–12. Only two other countries were allowed to increase emissions; these were Iceland (a 10 per cent increase) and Norway (a 1 per cent increase). Australia argued that its increase was justified for a host of reasons including the emissions intensity of the Australian economy and a high population growth rate. However, by any reasonable measure Australia received lenient treatment, as a result of its hard self-interested negotiating posture at the UNFCCC, and the desire of the other delegations to secure a deal. Other similar economies such as Canada were required to make emission reductions (in Canada's case by 6 per cent). Nonetheless, many of those who believed that Australia was not negotiating in good faith did not want to risk undermining a global deal by calling the delegation's bluff.

The second major allowance won by Australia was the inclusion in the Kyoto Protocol of the so-called 'Australia Clause' on land use. This enabled emissions from land clearing to be included in Australia's total emissions in 1990, inflating the baseline above where it would otherwise be. As land clearing was beginning to be brought under control by State governments in the 1990s, this meant that Australia would be able to meet its Kyoto target relatively easily, by counting reductions that were going to happen anyway. In some cases, the expected decrease in land-clearing was simply a function of running out of forest to continue clearing. An implication of the Australia Clause was that major smokestack industries in Australia such as power generation, aluminium and cement production could continue increasing their emissions to make up for these supposed 'reductions'.

Senator Robert Hill was reportedly greeted by a standing ovation from his Cabinet colleagues when he returned to Australia, and the then Howard government initially boasted of the achievements of Kyoto. Following the election of President George W Bush in 2000, which saw the United States take a new hard-line position against any mandatory emissions cuts, the Australian government followed suit. On World Environment Day, 5 June 2002, Howard announced that Australia would not ratify the Kyoto Protocol, asserting that it would cause irreparable damage to the Australian economy, while at the same time paradoxically committing the government to meet Australia's Kyoto target.[4]

By refusing to ratify the Kyoto Protocol, the Howard government stalled progress on the treaty's entry into force. Australian businesses also missed opportunities to be involved in global emissions trading under the Clean Development Mechanism.[5] It was not until February 2005, when Russia ratified the Kyoto Protocol, that it attained the required ratifications from industrialised countries to enter into force and thus become legally binding.

When the Rudd Labor government was elected to office in 2007, its first official act was to ratify the Kyoto Protocol. The Prime Minister personally handed the instrument of ratification to the United Nations Secretary-General Ban Ki-Moon at the climate conference in Bali in December 2007, where he received a standing ovation. By ratifying the Kyoto Protocol Australia gained a seat at the table in the twin climate change regime negotiating 'tracks' – under the UNFCCC and that under the Kyoto Protocol. This bifurcation was necessary because of the failure of the United States to ratify the Kyoto Protocol, which remains to this day the only major economy not to ratify.

Also in 2007, in an effort to improve emissions data and to lay the groundwork for an emissions trading scheme, the Australian Parliament enacted the *National Greenhouse and Energy Reporting Act 2007* (Cth) which established a mandatory reporting system for companies in relation to their greenhouse gas emissions, backed up by tough monetary penalties.

C Australia and the Copenhagen Accord

Prime Minister Rudd was personally involved in driving negotiations as a 'friend of the chairman' at the Copenhagen climate conference in 2009. While Australia did not play the role of wrecker at Copenhagen, the most significant climate conference since Kyoto in 1997, nor did Australia advocate an environmentally effective or globally equitable outcome. In advance of Copenhagen Australia offered to increase its 2020 emissions reduction target to a conditional 25 per cent reduction (below 2000 levels) if the international community agreed to global deal that would stabilise greenhouse gas levels at 450 ppmv (parts per million by volume, a measurement of the concentration of a gas).

When at Copenhagen, Australia advocated a new, legally binding treaty to reduce emissions that would replace the Kyoto Protocol and bring developed and developing states under the same regime.[6] An alternative proposal of Australia was for an amended Kyoto Protocol with new emissions reduction targets for existing parties, and an entirely new Protocol under the UNFCCC that would include mitigation targets for states including the United States and developing countries that do not have obligations under Kyoto. Australia's preferred position therefore paid little regard for the developed world's historical responsibility for climate change. It did, however, acknowledge the reality that, unless the major emerging economies of China and India curb their emissions, there is little chance of averting dangerous climate change.

Australia was one of 29 states parties engaged in the negotiations during the dying moments of the Copenhagen conference that produced the Copenhagen Accord. The Accord is a short document (less than three pages) and is a political declaration rather than a legally binding successor to the Kyoto Protocol. It allows developed and developing states to set their own emissions reduction targets and have them recorded by the UNFCCC secretariat.

The Australian government defended the outcome at Copenhagen, emphasising in particular the headline achievements of the conference – an agreement that increases in global temperature had to be limited to no more than 2°C, and that developed *and* developing countries needed to commit to limit emissions. Australia was also pleased with the inclusion in the Accord of a call for the immediate establishment of a mechanism to reduce emissions from deforestation and degradation (REDD).

In January 2010 Australia submitted its target range to the Copenhagen Accord. This simply restated Australia's pre-Copenhagen emission reduction targets – 5 per cent unconditional, with up to 15 per cent and 25 per cent both conditional on the extent of action by other states.[7] The government indicated that the 5 per cent reduction would not be increased until:

> [T]he level of global ambition becomes sufficiently clear, including both the specific targets of advanced economies and the verifiable emissions reduction actions of China and India; the credibility of those commitments and actions is established, for example, by way of a robust global agreement at the next United Nations Climate Change Conference in Mexico, or commitment to verifiable domestic action on the part of major emitters including the United States, India and China; and there is clarity on the assumptions for emissions accounting and access to markets.[8]

D Australia at the Cancún meeting in 2010

The climate talks at Cancún in December 2010 were more productive than those at Copenhagen. They marked an important return to the multilateral United Nations process and resulted in a package of decisions (the 'Cancún Agreements') formally adopted under the UNFCCC and the Kyoto Protocol (by contrast the Copenhagen Accord had only been 'noted' by the parties at Copenhagen.) The Cancún Agreements were supported by all parties to the UNFCCC and the Kyoto Protocol with the sole exception of Bolivia, and this high level of support was made possible by the deft diplomacy of the Mexican chair of the conference. Being decisions of the Conference of Parties to the UNFCCC and the Meeting of Parties to the Kyoto Protocol, the Cancún Agreements are regarded by most states as being more legitimate than the Copenhagen Accord, even though much of their core content is based on the Accord.

The Cancún Agreements under the UNFCCC negotiation track included a decision to hold temperature increases to below 2°C, with this goal to be kept under periodic review, and with the potential to lower the increase to 1.5°C. As regards mitigation, the parties agreed, among other things, to improve the reporting, monitoring and verification of the mitigation contributions made by developed and developing countries. In relation to reducing emissions from land-use change, the parties agreed that all states should aim to slow, halt and reverse forest cover and carbon loss, with developed states to assist through a REDD scheme which could involve carbon credits for avoided deforestation. The parties also agreed that more needs to be done to support adaptation efforts, especially in the most vulnerable developing countries.

Under the Kyoto Track, agreement was not reached on a second commitment period for the Protocol, which expires in 2012. However, the groundwork was laid for such an agreement at the next global climate talks in Durban in 2011. This included a recognition that, to prevent the most serious effects of climate change, developed countries would need collectively to reduce their emissions by between 25 and 40 per cent on 1990 levels by 2020; that emissions trading, the Clean Development Mechanism and Joint Implementation should continue; and that agreement on new targets was needed as soon as possible to avoid a hiatus following the expiry of the first commitment period.

Although the Rudd and Gillard governments have not played an active role as spoiler of international climate negotiations, Australia is far from being a leader in global efforts to confront the climate crisis. Australia has adopted a cautious 'wait and see' approach before making any unilateral commitment to reduce its carbon emissions at levels consistent with its share of responsibility for global warming.

E Australia's domestic policies on climate change

The election of the Rudd Labor government in 2007 raised expectations that Australia would take stronger action on climate change than the previous government had done. In opposition, Rudd (like his predecessor, Kim Beazley) pledged to develop a national emissions trading scheme, in addition to pursuing clean coal technology and mandatory renewable energy and energy efficiency targets. At his Parliament House climate summit in March 2007, Rudd even called climate change the 'great moral challenge of our generation'. He also announced the formation of the Garnaut Review on climate change and his plan to create a Department of Climate Change, with Senator Penny Wong as its first Minister.

The idea for an Australian emissions trading scheme (ETS) had been raised by the Australian Greenhouse Office in 1999, but the then Howard government did not pursue it. A proposal in 2003 for a national ETS was rejected by Howard's Cabinet, despite being supported by Treasury and various government departments.[9] Between 2004 and 2007 the Howard government sent ambivalent signals about the need for an ETS, opposing it on economic grounds yet suggesting that Australia might participate in one should the right kind of global system emerge.

In the meantime, at least one Australian State decided to act on its own initiative. In 2003, New South Wales introduced the Greenhouse Gas Reduction Scheme, one of the world's first mandatory emissions trading schemes.[10] In 2009, Victoria introduced an Energy Efficiency Target scheme, which places a liability on large energy retailers to surrender a certain number of energy efficiency certificates annually.[11] Neither scheme is linked to international carbon trading, though a federal scheme could enable this.[12] Some other Australian States rely instead on incentives to reduce emissions, as under South Australia's Residential Energy Efficiency Scheme, which encourages households to decrease their energy consumption.[13]

As for Howard's federal government, faced with growing public discontent over its climate policy, and under pressure from a Labor Opposition promising to sign Kyoto and create an ETS, in late 2006 Howard established the Prime Ministerial Task Group on Emissions Trading. Its report in May 2007 recommended the creation of an Australian 'cap and trade' ETS by 2012, which the government endorsed in July 2007. Criticisms of the details of that proposal were overtaken by the election of the Rudd Labor government later in that year.

At the election, neither party committed to specific emissions mitigation targets in the short term,[14] although Labor proposed a 60 per cent reduction on 1990 levels by far-off 2050. The new Rudd government still did not commit to such targets at the Bali conference in 2007, though it did not oppose a reference to the recommendation by the Intergovernmental Panel on Climate

Change (IPCC) to reduce developed country emissions by 25–40 per cent on 1990 levels by 2020, in order to keep warming below 2°C.[15] To put those figures in context, it is forecast that under a 'business as usual' scenario – where Australia does not reduce its emissions – by 2020 Australia's emissions would grow by 28 per cent on 1990 levels, or around 1.2 per cent per year.[16]

1 Rudd's Proposed Carbon Pollution Reduction Scheme

The Rudd government's effort to establish a national ETS was fraught with difficulty. In May 2008, the government proposed creating one by 2010. Its proposed Carbon Pollution Reduction Scheme (CPRS), foreshadowed in a Green Paper of July 2008, made various concessions to special industry groups, including forestry, agriculture, transport, coal power producers, trade exposed industries, and other affected businesses and communities.

In part, the government wished to placate industry groups which might otherwise oppose the CPRS, and to attract enough Opposition support in the Senate to pass the law. But in doing so, the government alienated others (such as environmentalists and the Greens Party in the Senate), particularly as more industry concessions were made during 2008–09 in negotiations with the then Leader of the Opposition, Malcolm Turnbull, to seek his support. Delays to the introduction of the CPRS were also agreed to, in part due to concerns about the global financial crisis.

The increasingly compromised scheme undermined the integrity of the CPRS (see further below at Part H). The Garnaut Review in September 2008 recommended establishing an environmentally and economically 'pure' ETS, with no free permits and limited concessions for industry groups. As the government increasingly revised the CPRS and departed from the Garnaut model, its perceived capacity to reduce Australia's emissions – and its economic viability – were both undermined, even as certain industries continued to call for further concessions. Issues concerning the purposes and optimal design of an ETS are considered later in this chapter.

The revised CPRS legislation was defeated in the Senate in August 2009, after failing to secure Opposition support. Keen to avoid a double dissolution election over the issue, Turnbull negotiated further changes to it with the government and eventually agreed to support it in late November 2009. The deal split the Coalition, a majority of whom opposed it, and Turnbull was soon deposed as leader by Tony Abbott, a climate change sceptic who had previously called climate change 'absolute crap'. Abbott rejected an ETS altogether, calling instead for an incentive-based (but not obligatory) emissions reduction fund and 'direct action' measures on solar power and soil carbon sequestration. The government's new CRPS Bill was defeated in the Senate in early December 2009, opposed by both the Coalition and the Greens.

Widespread disappointment with the Copenhagen conference, Abbott's aggressive labelling of the CPRS as a 'great big new tax on everything' and a looming federal election combined to spook the Rudd government in April 2010 into suspending the introduction of a CPRS until at least late 2012. The government's related policies - the Home Insulation Program, and the Green Loans Scheme - experienced serious administrative problems and were eventually terminated.

The failure of the CPRS - perhaps Rudd's chief mission in government - stoked a crisis of confidence in his leadership, resulting in his replacement as Prime Minister by Julia Gillard in late June 2010. Climate change was then taken off the agenda at the August 2010 federal election, in which Labor narrowly retained power by forming a minority government. Ross Garnaut described the protracted and messy CPRS policy process as 'one of the worst ... we have seen on major issues in Australia'.[17] It also contributed to a damaging loss of public faith in the government's response to climate change.

2 Gillard's 'Clean Energy Future' and carbon pricing

After the 2010 election, the Gillard government responded to the widespread disillusionment by establishing a Multi-Party Climate Change Committee in late September 2010, tasked to explore options for 'pricing carbon' and building community consensus for action. The Committee comprised members from the government, key independent parliamentarians and the Greens. The government invited the Coalition to participate but the Opposition declined.

The starting point for the Committee's work was that 'a carbon pricing mechanism is the most cost-effective and economically responsible way of reducing Australia's carbon pollution'.[18] The idea is that putting a price on carbon would encourage businesses to reduce their carbon pollution and to invest in clean (or cleaner) energy. As a market-based mechanism, it is also seen as the cheapest and most efficient way of reducing carbon emissions overall.

Despite criticisms that creating the Committee would simply delay action or result in a 'talking shop', in February 2011 the government endorsed the Committee's proposed strategy on carbon pricing. In July 2010 it then announced its plan for a 'Clean Energy Future', which aims to cut carbon pollution by at least 5 per cent compared with 2000 levels by 2020.[19] The legislation to create the scheme was introduced into Parliament in September 2011, involving 18 Bills. The Bills passed the Senate on 8 November 2011 and the *Clean Energy Act 2011* received the Royal Assent on 18 November 2011.

The two-stage plan commences on 1 July 2012 with a fixed price on carbon for three years, then converts into a 'cap and trade' emissions trading scheme. The carbon price would start at $23 per tonne and rise by 2.5 per cent per year over the transition period, moving to a market-based price under the emissions trading scheme. The carbon price will initially affect 500 of Australia's 'biggest polluters', who will need to buy and surrender a permit for each tonne of carbon pollution produced. It serves as an incentive for businesses to cut their pollution and find more efficient ways of operating. The transition period would give businesses time to adapt. Businesses would also be able to purchase permits through international carbon trading markets.

The revenue raised through carbon pricing would be used to subsidise households for increases in energy and costs of living, including through income tax cuts and increased family payments and pensions. In addition, revenue would be used to support jobs and competitiveness. An Energy Security Fund will also provide assistance to electricity generators most affected by the carbon price over the first five years, through free allocation of carbon permits and cash, in return for adopting clean energy investment plans. Billions of dollars of financial support for business innovation in renewable energy and low-pollution energy technologies would also be provided through a range of schemes,[20] which will help Australia meet its legislated Renewable Energy Target (where 20 per cent of electricity must come from renewable sources by 2020).[21]

The reaction to carbon pricing

Following the announcement, the government's approval rating appeared to plummet in one opinion poll, suggesting public dissatisfaction. The Opposition painted the proposal as a new 'carbon tax', and the government's awkward explanation of a complex scheme did little to dampen fears about a higher cost of living. There was confusion over many of the details, including the hard choices about which industries would be covered or exempted, the level of social compensation and what costs the community would have to bear.

The plan was most heavily criticised on the grounds that it would increase the cost of living for households and impose undue costs on businesses and the economy. The Opposition claimed, for instance, that there would be an additional $863 burden on families per year,[22] with increases in electricity, gas and transport.[23] It also claimed that regional areas would suffer because most jobs in mining, manufacturing and electricity generation are based there.[24]

The Federal Treasury estimated that the total increase in costs would be a more modest $510 per year.[25] It found that the carbon price's impact on everyday living expenses would be negligible (for example, the cost of milk would rise by less than one cent per litre, and rent would cost an extra 40 cents per week).[26] But consumer electricity prices would increase significantly by $3.30 a week, or $239 a year, accounting for much of the overall increase.

While these increases are substantial, the majority of households would receive compensation to offset them, funded by the carbon pricing scheme.[27] The government estimates that, while weekly costs would increase for the average household by $9.90, the average household would receive $10.10 in extra government assistance. Most households would be better off, and only high-income earners (over $150,000 per year) would be worse off, and they are more able to afford it.

Some economists have even suggested that the compensation is too generous,[28] with around $15 billion raised from carbon pricing being redistributed to households. Politically, it is understandable that governments are keen not to be seen to be increasing the cost of living. There is also a social imperative to ensure that people on low incomes are not adversely affected. But there is a danger in encouraging *middle class* voters to believe that they should be free to keep polluting without bearing financial responsibility for mitigating it. The exclusion of fuel from a carbon price is a case in point, leaving car pollution untouched. However, for all other household emissions sources the assistance is not related to a household's actual emissions and does not change the incentives to switch to cleaner alternatives.

The government discovered that on climate change, it was 'damned if it did, and damned if it didn't'. Former Prime Minister Kevin Rudd's support collapsed partly because he did too little on climate change, yet Prime Minister Gillard suffered because she did act. While the public generally supports the idea of addressing climate change, support begins to waver once people realise they may have to pay for it, when the details are uncertain and when political critics oppose it no matter what. The political and policy challenge for government is acutely difficult, particularly in the absence of bipartisan support for change.

Table 4.1: Overview of Carbon Pricing

Price of carbon emissions per tonne from 1 July 2012	$23
Market-based emissions trading commences	July 2015
The number of big polluters who will pay the carbon tax	500
Extra weekly costs for the average household	$9.90
Average weekly assistance for households	$10.10
Total compensation to households over first three years	$15 billion
Annual income at which compensation cuts off	$150,000
Target for carbon reduction by 2050, from 2000 levels	80%
Estimated reduction in carbon emissions by 2020	159 million tonnes

There has also been controversy over the impact of carbon pricing on Australian industries, with concerns from lobbyists that it will impose a burden that overseas competitors do not face.[29] The Opposition suggested that increased costs for rail, marine and heavy vehicle operators will adversely affect exports from regional Australia and the price of goods entering regional towns.[30] Small businesses will experience increased electricity prices but receive little assistance, leading to loss of jobs and increase in prices.[31] Export industries are said to face particular difficulties,[32] including risks of 'leakage' of manufacturing to overseas.[33]

Some carbon-intensive Australian States have also expressed concern about the special impacts on their industries and residents. The Victorian Premier, Ted Baillieu, released a report by Deloitte Access Economics that claimed the carbon price would cost Victorian households $1050 per year. It would also decrease the State budget by $660 million by 2015 and cost 35,000 jobs, with Gippsland, Bendigo and Barwon being especially hard hit.[34]

Similarly, modelling by the Bligh government showed that Queensland would be hit harder than other States, costing 21,000 jobs and reducing economic growth by 2.76 per cent by 2020. In contrast, federal Treasury modelling suggested that economic growth would slow by only 0.4 per cent.[35] The NSW Premier, Barry O'Farrell, also announced that a carbon price would cost its economy $3.7 billion and 31,000 jobs, with the Hunter, Illawarra and Central West most affected, along with the coal and aluminium industries.[36]

Others have argued, however, that these estimates do not account for $5.5 billion in compensation to the electricity industry (most of which is slated for coal power generators), a $300 million package for the steel industry, $10 billion for renewable energy projects,[37] and innovation driving new business and employment opportunities in a cleaner economy. Access Economics estimates that a more modest 28,000 jobs would be lost in regional Australia.[38]

More generally, the independent Grattan Institute argued in 2011 that the compensation for industry is too generous and will artificially prop up dirty or inefficient companies which do not deserve to survive in a cleaner economic future.[39] It found that, even with a higher carbon price and no compensation, mines would remain open and profitable, and liquefied natural gas operations could continue without being adversely affected. It saw the compensation package as an 'unjustified windfall gain' that unduly impaired the environmental benefits of the tax.

Although the Institute acknowledged the reasonableness of some free permits for the steel industry (where profitability is already poor), it stated that the current compensation scheme would provide more assistance to steel mills than they actually pay out for carbon permits in the first four years, rendering the scheme ineffective in reducing emissions in that sector. Pressure from industry lobby groups has undermined the environmental effectiveness of the scheme, as government responds to special pleading which sometimes exaggerates the costs for business.

Much of the criticism of the carbon price on economic grounds has been inaccurate, deliberately misleading or has represented special interests rather than the national interest. In fact, economic modelling by Treasury predicts that, even with a carbon price, employment would grow by 1.6 million jobs by 2020.[40] Access Economics suggests that employment will grow even in coal-intensive regions such as the Hunter Valley in NSW or Gippsland in Victoria.[41] One of the reasons for new employment opportunities is precisely *because of* carbon pricing – the transition to a 'cleaner' economy will drive investment and jobs in new technologies and industries, which will help to offset losses in the older, 'dirtier' economy. Australia risks missing out on these new opportunities if it delays its transition; already, other countries are gaining a competitive advantage over Australia in the new technologies and industries, including in exports. Of course, it goes without saying that the costs of global inaction will also have profound economic costs, as Chapters 1 and 2 have demonstrated, and this too must be factored into any analysis of a carbon price and any alternatives to it.

Despite the exaggerated criticism of the costs of the government's scheme, the Coalition in Opposition's 'Direct Action Plan' would likely be more expensive and less effective. The plan involves the creation of a general pool of funding, sourced from savings to the budget, which would be distributed by the market towards the lowest cost projects. It uses incentives rather than penalties and relies on the government redistributing tax revenues to businesses rather than creating an economically viable market mechanism for carbon trading.

While incentives can induce behavioural change, in the absence of penalties no price signal would flow to consumers to drive the demand side of abatement,[42] encouraging consumers to continue polluting as usual. Direct action also forgoes economic opportunities for linking in with international abatement schemes. The federal Treasury estimates that the cost of direct action would be $62 per tonne of carbon emitted, reducing Australia's GDP by 1 per cent by 2010 (compared with $29 per tonne and 0.5 per cent of GDP under the government's scheme).

The rest of this chapter steps back from the details of the carbon price debate and considers more broadly the policy options open to Australia in confronting climate change. It explores the advantages and disadvantages of a carbon 'tax' versus an emissions trading scheme, and of different kinds of emissions trading schemes (such as 'cap and trade' or 'baseline and credit'). It shows that each of those options can also be designed in different ways (as the above discussion of the 'Clean Energy Future' plan suggests) and that certain policy choices (such as exempting or subsidising certain polluters) can undermine the environmental effectiveness and economic efficiency of a given scheme – even if it makes it more politically palatable. Before turning to those issues, the next section discusses whether Australia should act now on climate change or wait for the international community to respond.

F Should Australia act now or later?

Given the slow pace of international cooperation since Copenhagen, and the political obstacles encountered in Australia, it is tempting to think that Australia can afford to wait before taking real action on climate change. Australia currently has only around 22 million people and is therefore responsible for only a small fraction of total global emissions, meaning that Australia plainly cannot solve climate change alone, no matter how far it reduces its emissions. With many other countries delaying action on regulating carbon emissions, an understandable question is why should Australia act before others, especially when it is competing in a global marketplace and risks being economically disadvantaged by doing so?

1 Acting before developing countries

There are compelling reasons why Australia should act before the developing countries, including China and many of our regional neighbours. These reasons are grounded in a basic notion of fairness: Australia's historical contribution to the problem has been much greater, its present emissions are significantly higher and it has greater wealth available to invest in solutions. As noted earlier, Australia contributes 1.5 per cent of the greenhouse gases emitted worldwide each year.[43] This is smaller than countries such as China (16 per cent), the United States (16 per cent) and Russia (5 per cent), but only because our population is much smaller. Per capita, as the table below shows, Australia's emissions are among the highest in the world (depending on the year, often in first place among the advanced economies).[44] Each Australian has roughly four times the carbon footprint of the average Chinese person. A per capita measure (see Chart 4.1) is the only fair way to assess the impact of different nations.

Yet, these notions of fairness quickly translate into self-interested practicalities. Since Australia cannot force its neighbours to accept an outcome that is too egregiously unfair, Australia is very unlikely to get a workable solution unless it acts before the developed countries. However successful Australia is at convincing itself that China and the developing world must 'pull their weight' in the short term, Australia is unlikely to succeed in convincing them that they must take actions that Australia itself has not yet taken to clean up a global problem that it has caused. This would hold true even if those countries had the same economic and living standards as Australia. In fact, however, they suffer from extensive poverty, which makes it harder to demand that they reduce their emissions earlier in their stage of development than Australia did.

Nonetheless, China and other developing nations have demonstrated their seriousness on this topic (as discussed in Chapter 3), including significant reduction actions, setting up necessary reporting systems and undertaking voluntary emissions targets

Chart 4.1: Per capita emissions of major countries, 2005 (metric tonnes)

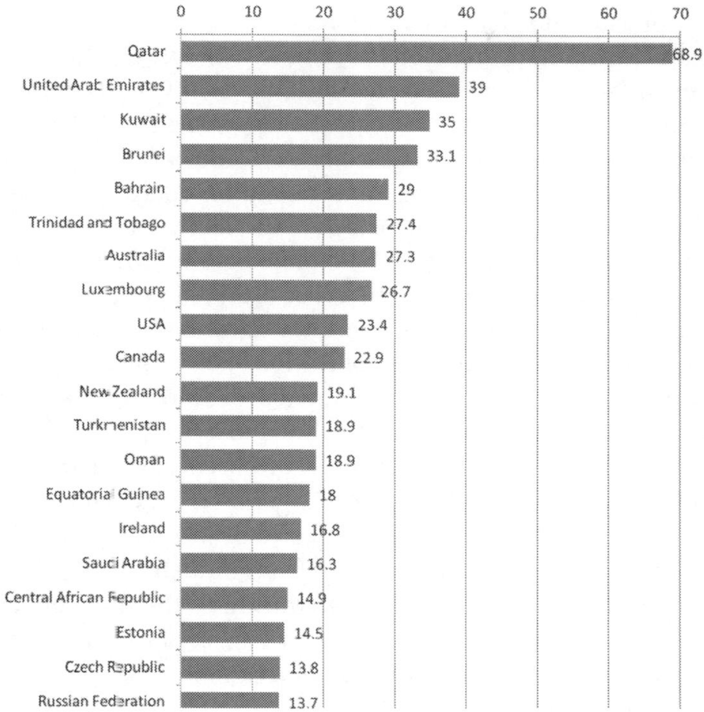

Source: World Resources Institute (2005), available at <http://www.wri.org>

and commitments. There is no reason to expect that they will reject fair commitments going forward. This is especially so given that strong direct incentives can be set up for them to do this, such as developed countries choosing to impose carbon tariffs on exports from non-compliant countries.

2 Moving before other developed countries

The case for moving ahead of other developed countries is less strong, given the relatively limited number of comprehensive emissions trading schemes elsewhere, but is still no reason to delay action. The 27 member states of the European Union, for instance, have been well ahead of Australia, committing in 2007 to reduce emissions by 20 per cent of 1990 levels by 2020, and increasing to 30 per cent if other countries committed to a fair global climate deal. The European Union has also had mandated limits on carbon emissions and functioning emissions trading scheme since 2005 (which now includes an additional three non-European Union countries - Iceland, Liechtenstein and Norway). Europe's carbon markets will soon be entering the next phase, expanding into new sectors (such as airlines and petrochemicals) and taking on board the lessons of the initial rocky roll-out, which gave too many free permits to industry. Spain, Germany and the Nordic countries have mature renewable energy industries with major production capacity, and wind turbines are a relatively common sight in the countryside throughout the major economies of the European Union.

Critics of the Gillard government's carbon price scheme suggested that it would commit Australia to a far greater burden than that assumed by EU countries - supposedly being 18 times larger in dollar terms, and 400 times larger on a per capita basis.[45] Television advertisements by industry groups in 2011 claimed that it would strip 14 times more money from industry in Australia than under the EU scheme.[46] Yet, according to the independent Climate Institute, those claims are misleading, because they ignore Australia's free permits for businesses, and significantly undervalue the costs imposed by the EU scheme.[47] As Professor Ross Garnaut observes, even without comprehensive emissions trading schemes, other regulatory processes in countries like the United States and China have imposed more costs on businesses and households, to achieve equivalent emissions reductions,[48] than the relatively more efficient Australian proposal. It is true that Australia's scheme would be one of the more ambitious ones. But some Australian industries are simply campaigning for the right to continue polluting, when the uncompensated costs of climate change harm will be borne by everyone.

Elsewhere, Japan took on its Kyoto commitments following an already significant energy efficiency drive in the 1980s that made further reductions especially challenging. Japan has been an innovator in financing carbon reductions offshore through carbon trading mechanisms. By contrast, legislative proposals in the

United States have been unable to secure enough support. Federal government policy in the United States nonetheless calls for steeper long-term cuts (80 per cent by 2050) than Australia (80 per cent by 2050). Individual States within the United States (such as California and nine other States) have been taking strong measures to reduce emissions.

Despite over a decade of engagement and negotiation at the global level on the topic, greenhouse gases can currently be emitted in Australia in unlimited quantities by any industry or individual who chooses to do so. If Australia does not reduce its emissions, it will remain a 'free rider', sponging on the good faith efforts by many other countries to reduce global emissions. If Australia chooses to curb emissions before the United States does, this may increase pressure on that country to come into line with international efforts – just as it did when it became the second last country to ratify the Kyoto Protocol in 2007, leaving the United States alone among all major United Nations member states in failing to join it.

3 Affordability for Australia

Ultimately, if everyone waits for someone else to act first, everyone will be worse off – a classic tragedy of the commons. Since the costs of climate change may far exceed the costs of reducing emissions – at least in the long term – it makes sense to find a way to avoid this outcome. Economic analyses indicate that Australia can easily afford to reduce its overall emissions in line with the 20–40 per cent cuts below 1990 levels by 2020 identified by the IPCC as necessary for developed countries. A reduction in the middle of this range (30 per cent) is estimated to cost around $2.9 billion per year – or $290 per household per year – amounting to merely 0.2 per cent of GDP.[49] Deeper reductions of 60 per cent below 1990 levels over a slightly longer period to 2030 would cost $590 per household, or around 0.35 per cent of GDP.

Annual household incomes in Australia are projected to rise significantly by 2020, so these costs are expected to be compensated for by much larger concurrent overall growth in national wealth and income. The Garnaut Review, for instance, estimated that it would cost the Australian economy around 0.1 to 0.2 per cent of annual growth to reduce emissions. These projections are also somewhat cheaper than the estimated global burden of mitigation. The Stern Review, for instance, found that stabilising atmospheric greenhouse gases at 500–550 ppmv by 2050 would cost 1 per cent of annual global GDP.[50] That suggests that, relatively, Australia may be better placed to respond than many other countries - particularly in the developing world - and indeed is well situated to contribute to global climate financing (for mitigation and adaptation).

To put these costs into perspective, recall from Chapter 2 the economic costs of *not* acting to reduce carbon emissions. These are estimated to be in the order of 3 per cent of global economic output with a 2–3°C rise and up to a 10 per cent decline in global GDP with a 5–6°C rise (the difficulty of estimating these costs, and potential for damage to be far worse than suggested by these figures, was discussed in Chapter 2).[51] Australia would not only be affected by the costs of climate change impacting directly on Australian territory, but also by the costs of responding to its impacts on other countries in the region.

In particular, since the bulk of Australia's trade is with Asian countries, downturns in their economies due to climate change impacts, and decreasing commodity prices, would undermine the value of Australian exports and the industries and employment they support.[52] At the same time, Australian foreign aid, development and humanitarian relief assistance, and national security resources, would be increasingly burdened by the additional needs of our neighbours.[53]

G In which areas can Australia reduce its emissions?

As a developed country, Australia is among those countries that have committed in international agreements to move first. It must make sufficient emissions reductions to contain the problem, while recognising that other, poorer countries' emissions will continue to grow in the short and medium term, and Australia's reductions will need to make up for that. But among the developed countries, Australia is more dependent on fossil fuels than most and has used this to negotiate a less onerous reduction burden in international climate change negotiations to date.

The basis for this lighter burden can be debated, however. Detailed economic studies indicate that the flip side of Australia's higher per capita emissions is that a lot of 'low hanging fruit' exists, allowing emissions reductions here that are cheaper - not more expensive – than in Europe or the United States. This is undoubtedly true in the power sector, as discussed below, which is Australia's largest source of emissions.

Ideally, emissions reductions would be made in those sectors of the economy where they can take place most cheaply. Any other outcome means that Australia is spending more of its resources than is absolutely necessary. Some sectors of the economy will need to undertake proportionally greater cuts than others - for example, the building and transport sectors have particularly cheap emissions reduction opportunities, along with heavy industry and agriculture. Exemptions in one sector increase the burden on the rest of the economy and the need to employ more expensive measures to achieve the same reductions. In some cases, such as for emissions intensive trade-exposed industries, there are reasons why this should still be considered in the short term, as discussed below. But any considerations of fairness between regions or sectors are better addressed through cash compensation rather than by shifting the burden of the reductions, since this preserves transparency as well as the incentive to reduce emissions across the economy.

Chart 4.2 (*over page*) sets out the major sources of Australia's carbon emissions by sector. The following section then explores how emissions can be reduced in each of the main sectors.

1 Energy

The electricity sector is by far the largest source of emissions in Australia, representing 35 per cent of total emissions.[54] As a result, reductions in this sector must be the top priority if national reduction targets are to be met. Fortunately, this sector also presents comparatively cost-effective opportunities. Reduction can occur both by changing the sources of energy and by improving patterns of energy usage and efficiencies.

Australia has one of the most carbon-intensive power sectors of all developed countries due to its overwhelming reliance on

Chart 4.2: CO_2 Emissions in Australia by sector, 2010 (%)

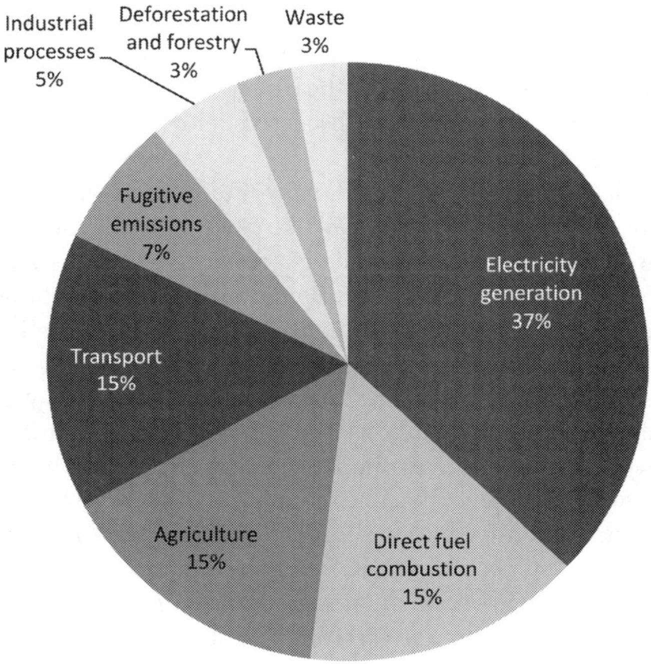

Source: Australian Department of Climate Change and Energy Efficiency, National Greenhouse Gas Inventory 2011.

coal. Other comparable economies either have significant nuclear power generating capacity, such as the United States and Western Europe, or a greater reliance on natural gas, as in Russia. As a result, emission reductions in this sector are actually cheaper for Australia than for these other countries - in contrast to common misconceptions. Each megawatt of renewable energy capacity constructed in Australia (such as wind turbines) displaces a much dirtier mix of generation technologies than it does in most other places, such as Europe, where it may simply displace an already low carbon fuel such as nuclear energy. This translates to more carbon reduction per dollar invested or, in other words, a lower cost of emissions reduction. Chart 4.3 opposite shows Australia's current electricity energy mix.

Chart 4.3: Australian electricity generation by fuel source, 2008–09 (%)

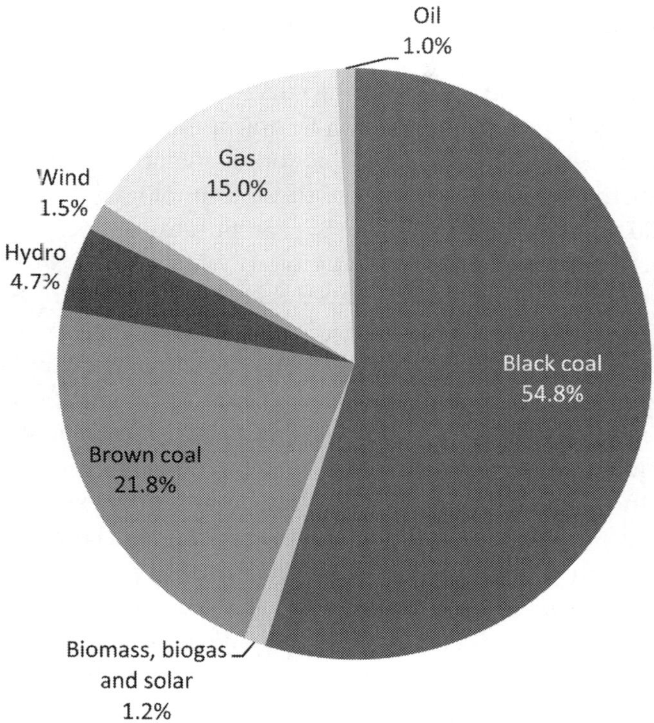

Source: Australian Bureau of Agricultural and Resource Economics and Sciences, Energy Update 2011.

It is estimated that, by 2020, 93 Mt of reductions per year can be achieved in the energy sector at an average cost of $55 per tonne, under a scenario consistent with a national reduction of 30 per cent below 1990 levels. This reflects the cost of choosing to meet demand growth through new renewable energy technologies, such as wind, geothermal and solar, rather than through building new coal-fired power plants, but does not require shutting down existing assets before their lifetime is up. Opportunities in the power sector tend to be more expensive than many energy efficiency measures, but can be more easily implemented through a smaller number of larger projects.

The most economically important renewable energy technologies at present are wind, solar and geothermal power (using hot rocks to heat water that drives a turbine). Wind is currently the most cost effective, mature and scalable, but solar and geothermal power are on a steeper 'learning curve' – meaning that their costs are declining more quickly as technical and business advances lead to new methods of production, installation and operation. Predicting the relative costs of these technologies over the next 10–20 years is impossible, but if present rates of decline in fossil fuels continue, there are scenarios in which renewable sources of energy could become cheaper than conventional ones. Many experts expect that some premium cost for renewables will remain through this time-scale – however, there is no underlying law that makes this necessary. In fact, renewables must by definition become cheaper eventually, since non-renewable fuels become scarcer as they continue to be used.

In addition to renewables, two non-renewable sources of energy are also important for emissions reduction because of their lower carbon intensity: natural gas and nuclear power. Depending on its source, gas can have roughly half the carbon intensity of coal per unit of energy, and the only carbon emissions attributable to nuclear power are from indirect activities, such as mining the fuel, and building and decommissioning the assets – many of which apply to any technology.

Renewable energy – primarily sourced from wind farms – can currently be purchased in Australia by consumers at an additional cost to conventional power. Supplies of wind and sun in the country are much greater than Australia is ever likely to need for power consumption. So what stops Australia deploying more wind power for all its needs, assuming it is prepared to pay the premium? In the short term, the answer is that wind is an intermittent power source, generating power only when the wind is available. Solar power behaves similarly but geothermal and hydro power can generate constantly (as can the low carbon non-renewables, gas and nuclear). Because a typical coal plant can generate power around 90 per cent of the time, our power consumption patterns have evolved to take advantage of this, with large industrial loads like aluminium smelters engineered to use technologies that provide a constant 'base-load', taking advantage of discounts offered by energy producers.

With no changes to existing energy distribution and use patterns, at least 20 per cent of Australia's power can be generated by intermittent renewable sources without leading to an increased risk of blackouts. It need not stop there, however. Three solutions are possible for unlocking the full potential of renewable power in Australia. First, there needs to be a shift away from base-load industrial technologies toward technologies whose demand for power can more easily be shifted to match supply. For example, electric arc furnaces – which have additional economic benefits – can increasingly be built instead of blast furnaces for metals production. Appropriate price signals can bring about these shifts, just as the discounts offered to base-load users led to our current patterns of demand.

Secondly, 'smart grid' technologies – that is, information processing and communication technologies – can enable much greater coordination between supply, demand and distribution in our energy grid, with the side benefit of also eliminating significant sources of waste and unreliability. For example, using smart grid technologies, pool pumps, refrigerator compressors and dishwashers can time their operation to avoid periods of peak demand or low supply, and power companies can avoid the costs and emissions in having substantial excess capacity on standby for peak periods. Real-time feedback to consumers can encourage further savings by making energy waste more visible through in-home displays and alerts.

Thirdly, related to smart grids is storage. Energy storage technology is currently too expensive for extensive use as part of the national energy supply, but there are several opportunities for this situation to change. Storage of energy using large-scale batteries or other technologies can eliminate the intermittency from sources like wind and solar by smoothing over fluctuations. Electric vehicles also represent a large potential source of grid storage – if substantial numbers are plugged in to the smart grid when not in use, their batteries could be used to automatically store and release energy at times of peak demand, appropriately compensating their owners (this application is referred to as V2G, vehicle-to-grid).

2 Transport

Just over 14 per cent of Australia's emissions derive from transportation.[55] Ultimately, achieving major emissions reductions will require replacing our reliance on private, fossil-fuel burning vehicles with alternatives. Public transport, better urban design and bicycling are major opportunities, with significant health and public safety side benefits. To the extent that private vehicles continue to be used, however, they will need alternative forms of energy.

As at 2011, most major global automobile manufacturers have either announced or begun delivering electric cars. When charged using renewable sources of energy, these vehicles offer the opportunity to reduce the emissions associated with driving them to zero. The Nissan Leaf, Renault Fluence and Chevrolet Volt are three prominent models currently being delivered in major markets in the United States and Europe. Smaller start-up companies, including Tesla and Coda, have further models available or announced. These cars are expected to be available in Australia within one to two years. Even when charged with a mix of non-renewable sources, the engineering efficiencies of electric motors mean that, in many cases, electric vehicles still reduce emissions - but their full potential is unlocked as the energy supply is decarbonised.

It is likely to take several decades before Australia's vehicle fleet could be converted to electricity, however, because of the lifetime of the average car and the currently limited quantities of electric cars being produced. For remaining internal-combustion vehicles, new generation bio-fuels may play an important role. In 2010, over one-third of all petrol sold in New South Wales contained 10 per cent ethanol from bio-fuels, with the possibility of further increases being limited by current supply shortages. While the production of bio-fuels itself requires energy inputs, bio-fuels can in principle reduce carbon emissions overall, although it is uncertain at this point how much is saved, depending on the type of bio-fuel and how it is produced.

Bio-fuel technology suffered from false starts, with first generation fuels diverting crops, arable land and scarce water from food production into fuel. The result was a spike in the prices of essential food grains and serious impacts on food security in developing countries. New generation bio-fuels may be able to avoid these problems by producing fuel from non-food plants (such as grasses), being grown in marginal land areas, and using less water than other crops. Research and development continue at pace into these technologies.

Fuel efficiency measures also help reduce emissions from fossil-fuel burning vehicles in the transport sector, involving incremental improvements in vehicle design (lower weights, better aerodynamics, engine control management, scrubbing exhaust fumes and hybrid engines).[56] These are necessary but interim measures, however, since the magnitude of reductions required exceeds the most optimistic fuel efficiency savings.

Greater investment in public transport – often not an historical strength of many Australian cities – can reduce Australia's heavy and inefficient reliance on car transport (especially single passenger journeys). Effective public transport offers the potential for significant economic gains due to its overall greater efficiency: more people moved, more quickly, using fewer assets, less energy, and a smaller proportion of scarce urban land. Cities like New York illustrate the potential here, with only a small minority of residents of the central borough of Manhattan owning vehicles, and a fast, effective 24-hour subway system with a single, low price for travel anywhere within the city. Cycle paths and incentives to use them are additional, important, contributions. Cycle paths have been used successfully in overseas cities, such as Copenhagen, where bicycle traffic often outnumbers and outpaces vehicle traffic on major streets.

The effectiveness of such investment is closely related to broader questions of urban design, particularly increasing residential densities and the location of work places and services. Getting those policy settings right in Australia's sprawling and poorly coordinated cities is also socially and economically important, quite apart from any carbon reduction benefits they may bring. Questions about urbanisation are, in turn, related to the size and sustainability of Australia's population.

3 Building, industry and mining

Almost 15 per cent of Australia's emissions come from building, heavy industry and mining.[57] As in the transport sector, there are a variety of readily available technologies that can assist in reducing emissions in these sectors.

In the construction of buildings, these include improvements in commercial air handling, air conditioning and residential water heating systems; better insulation; more efficient lighting; energy efficiency standards for appliances or motor-drive systems; and mine methane recovery or leakage prevention. All of these opportunities can reduce the energy consumption of buildings. Although energy consumption does not usually result in direct emissions from the buildings themselves, these represent indirect emissions reduction opportunities that translate into direct reductions in the emissions from power plants.

A range of industry-specific reduction opportunities exists in heavy industries. For example, substituting clinker - an energy-intensive input material used in cement manufacturing - with other materials can reduce emissions significantly. In petroleum production, reducing gas flaring by capturing rather than burning waste gases can represent similarly major emissions reductions.

4 Forestry and agriculture

Almost 16 per cent of Australia's emissions derive from agriculture and 4 per cent from forestry.[58] As noted earlier, Australia receives credit under the Kyoto framework for a large amount of easily available reductions through the restriction of land clearing. Beyond this, there is further potential for avoiding deforestation, and for reafforestation for commercial forests and plantations, as well as replanting of agriculturally marginal crop and grazing lands.

There are also readily available technologies and changes in farming practices which can assist in achieving substantial emissions reductions. These include zero-till cropping or changes in tillage (such as reducing ploughing to retain carbon in the soil); improved fertilisation techniques; improving livestock feeds to reduce methane; capturing methane from landfills; wider use of permaculture; productivity gains;[59] and treatment of manure and the use of nitrification inhibitors on pasture.

Something as seemingly simple as holistic management of grazing, involving better stock rotation and land management, can also have big impacts on reducing emissions, while even increasing stock numbers and not necessarily adding to overall workloads.[60] Diversification of agriculture production, rather than its concentration, can also bring benefits.[61] Investment in new infrastructure, though costly upfront, can also pay off, as in the use of pyrolysis, which involves heating biological matter without oxygen (such as crop waste, manure or forest off-cuts) to produce charcoal, and which generates energy, improves soil and reduces atmospheric carbon.[62] Australia's Carbon Farming Initiative allows for credits from actions which reduce or store pollution in the land sector to be sold to polluters under the carbon pricing mechanism. The government is also planning to reinvest 1.7 billion of carbon price revenue in land sector measures to unlock the abatement potential and support important productivity and biodiversity co-benefits.

5 Behavioural change

Most of the areas discussed above involve technological fixes or changes in sectoral behaviours. But it is also important to highlight that emissions can also be reduced through individual or group behavioural change, which can come about as a result of increased personal knowledge of how each of us contributes to climate change, or through financial incentives that follow from carbon pricing.

There are several large opportunities for feasible behaviour change to reduce carbon emissions. In buildings, large savings are possible through a reduction in air conditioning and heating use, water heating, and switching off unneeded lights and appliances. In transport, significant savings are possible through a greater use of public transport, walking and cycling, putting more passengers in one car, and choosing smaller cars. Reducing air travel (for example, through the use of teleconferencing) is also a major opportunity, since for those who fly regularly it is likely to be the largest source of individual emissions. In the agriculture sector, very significant quantities of emissions can be avoided by shifting our dietary habits.

Reducing meat consumption and substituting consumption of ruminants (beef and sheep) with other meat can enable significant reductions because of the intensity of carbon emissions associated with farming these animals.[63]

There are numerous small changes that individual Australians can easily make without much sacrifice or even inconvenience, yet which cumulatively can make a big difference to the country's overall emissions. Deeper cultural changes are harder to induce, particularly those which call for a scaling back of personal consumption. Yet, the fewer changes that are made now to mitigate emissions, the more drastic may be the impacts of climate change on Australia's standard of living in the future.

H The policy options for mitigating Australia's emissions

It is one thing to understand what activities must be undertaken to reduce emissions and another thing to develop the policy mechanisms to make those things systematically happen. For example, wind farms may need to be constructed, but who should construct them and how should they be paid for? Buildings may need to be fitted with energy efficiency features, trees planted and changes made to industrial production processes, but, in each case, some incentive or mandate must bring about these actions, with a minimum of waste and cost.

In a market economy, where most production decisions are undertaken by private firms that are independent of direct government control, three forms of intervention are possible. First, governments can act directly to carry out or contract the performance of a particular activity. Secondly, governments can mandate or compel certain firms or individuals to carry out the action. Thirdly, governments can impose taxes and liabilities and thereby create an incentive to carry out the action.

In general, governments in market economies want to minimise the extent to which economic activity is directly managed by their agencies, so the third option is preferred over the second, which in turn is preferred over the first. This allows the maximum degree of flexibility and innovation to take place, with the idea being that entrepreneurs and competition can bring about more efficient solutions than might be imagined in advance by a more prescriptive government agency. The assumptions behind these preferences are not interrogated here, but they are simply stated as a means of understanding emissions reduction choices within existing economic policy.

1 Direct government action

Although markets are useful instruments for efficiently organising many kinds of production, there are also many activities for which they are of limited (if any) utility. In these cases, there is a greater need for direct government action to enable emissions reductions. A clear example of this is the planning and implementation of major transport infrastructure. No incentive or mandate can bring about a shift from the use of private cars to trains if there are no trains in the first place or if the trains offer an inadequate service. The natural monopoly of a public transport system, the fact that it may require subsidies in order to operate solvently, and the long-term, large scale nature of the investment required mean that direct government action is generally required to roll out the infrastructure for a wide-scale shift away from private vehicle transport.

2 Mandated action

There is a broad class of mitigation actions that can most efficiently be carried out by individuals and firms, but for a number of reasons will not happen sufficiently even with clear economic incentives in place – that is, even where implementing them actually saves, rather than costs, money. Many energy efficiency opportunities fall into this category. For example, it has long been known that both cash and emissions can be saved through a range of upfront investments to reduce energy wastage in the heating, cooling and lighting of buildings. The use of compact fluorescent and LED lighting are key examples. Although an LED light bulb costs more upfront than a conventional one, its much longer lifetime and dramatically reduced energy usage means that it unambiguously saves money in the long run.

Simplistic views of economic behaviour led many to expect that these products would quickly become widespread. Yet after years of availability, people still buy the older, less efficient bulbs. This is often presented as irrationality or ignorance: why, after all, should consumers choose to waste both money and energy, when clear alternatives are available and promoted?

More careful consideration of the forces at work makes clear that there are several reasons for 'market failures' of this kind. One is a misalignment of incentives - for example, where an energy efficiency measure benefits a tenant (in lowering their energy costs) but not the builder (who must pay for the installation of energy efficiency measures, such as insulation, from which the builder derives no benefit). Another reason is that there may be a long pay-back period involved in reaping the benefits of a measure. For example, installing solar panels involves high upfront costs that may take a long time to recover through lower energy costs. People tend to discount these future savings significantly - and not always for irrational reasons. After all, who knows what might happen in 10 years' time? A person might move house, or have a substantially higher income, or experience some tragedy, any of which would make this upfront investment less attractive.

These kinds of market failures are not confined to environmental issues. Even when the risks are far more direct and personal, consumers exhibit the same tendency to discount benefits to their lives in other areas. Smoke alarms are a clear example of this - a basic familiarity with the cost and payoffs of installing one should be enough to ensure that they would be installed in every house and apartment. But after repeated failures by homeowners to install them and repeated unnecessary deaths, they were mandated. This is true not just in Australia but in most developed countries.

In these cases, clear, direct regulation of the underlying activities can bring about behaviour that is necessary and in the interests of consumers, but would not otherwise happen, at least not to the extent required. Incandescent light bulbs are now being phased out in Europe and Australia through government bans. Star rating systems for building energy efficiency have been introduced and minimum levels are progressively being ramped up, although a consistent national scheme and sufficiently strict standards are still lacking.

Clearly, the government cannot regulate rational behaviour in all areas, but when sufficiently large mitigation opportunities can be captured through mandates, with minimal inconvenience and while saving money for consumers, a strong case can be made. This is consistent with conventional regulatory approaches in other areas.

3 Taxes and liabilities: carbon tax or emissions trading?

The mechanism that has been preferred in recent years by most economists and the Australian government is to create a clear and transparent economic disincentive for emissions. This can be done by taxing goods and services proportionally to the emissions they generate (a carbon tax), or attaching other liabilities such as a requirement to purchase permits to cover the quantity of gases emitted (emissions trading). The cost of permits, in dollars per tonne of emissions, can either be set by the government or through trading on a regulated carbon trading market. Australia's 'Clean Energy Future' carbon price is a hybrid model, commencing with a fixed price (like a tax) for an initial period, before moving to a market-based price. In contrast, the abandoned CPRS proposal largely priced carbon through market trading.

Simple carbon taxes have been implemented in other countries, particularly in Europe and affecting fuel, petrol, gas, electricity and high-polluting vehicles. The United Kingdom, for instance, introduced a 'climate change levy' in 2001, increasing electricity costs by 15 per cent.[64] Trading in environmental products has previously occurred in areas such as acid rain, renewable energy, biodiversity[65] and now carbon. A 'cap and trade' emissions scheme sets a total emissions target and individual companies are allocated permits up to that cap (an example is the European Union's emissions trading scheme and the scheme under the Kyoto Protocol). A 'baseline and credit' scheme sets a baseline performance for each company (based on industry best practice or the company's past performance) and any emissions savings generate credits for the firm (an example is NSW's Greenhouse Gas Reduction Scheme). In both cases, surplus permits or credits can be sold or even banked towards future compliance and there may be penalties for exceeding emissions limits.

There is a fundamental difference between a carbon tax and an emissions trading scheme. Under a carbon tax, the government sets a price, enabling firms to make investment decisions without undue uncertainty over the cost of emissions. There is no predetermined limit on the country's total emissions, since anyone willing to pay the price can obtain permits. If the price is too low, or demand is higher than expected, there will be more emissions than expected and the country risks overshooting national targets. This can happen either because mitigation actions cost more than expected or the economy grows more strongly.

On the other hand, if a market is used to set the price of emissions by auctioning a fixed number of emissions permits consistent with national targets then, in theory, the total emissions will be fixed. But since the price floats, firms will not be able to predict precisely the price of those permits, leading to greater uncertainty around the business case for the new investments required. If those investments do not take place in time, the price of carbon permits will continue to rise until the government is forced to choose between bankrupting the businesses involved or undermining its carbon scheme.

Many investors can be expected to judge that the government will not force these large businesses into bankruptcy. This directly impacts on the credibility of the permits and, as a result, their price. The Australian government's earlier proposed CPRS capped the maximum price of carbon permits in advance, in effect conceding that if businesses do not make the necessary investments they will nonetheless be able to continue emitting carbon in excess of national targets.

In theory, an emissions trading scheme in which the permits to emit greenhouse gases are auctioned, prices are not capped, and where good information is available about the likely future cost of making emissions reductions, can have a better environmental and economic outcome than a carbon tax. This is because it can, in principle, set a fixed level of emissions and allocate those emissions rights to firms that are able to create the most value from them.

Introducing a carbon price changes the relative prices of goods and services, which, in turn, makes emissions-intensive goods more expensive than less emissions-intensive goods. This encourages consumers and businesses to change their behaviour (including through innovating to produce new services that emit less carbon) and ultimately to reduce emissions. Businesses will pay for permits if they are cheaper than the cost of reducing their emissions, and will reduce their emissions if doing so is cheaper than the price of the permits. Businesses will sell permits if doing so is more profitable than using them to produce goods and services. The idea is that market incentives are a better way of making the most valuable use of carbon emissions and of reducing emissions than relying on less sensitive government directives.

However, once free permits are given out to favoured industries, prices are capped, and other complexities introduced, the beneficial aspects of an emissions trading scheme become less clear. Some believe that permit prices will always vary unpredictably (just as many other commodities do), as they have done in Europe, no matter how the scheme is designed. This is where the predictability of a carbon tax, where the cost of carbon emissions can be factored into investment decisions with a much higher degree of certainty, offers real practical benefits. Such a scheme may be less efficient than a pure, well implemented ETS but will arguably be more effective than a loophole-laden, bad ETS.

4 Free permits?

Australia's earlier CPRS proposal, and the later 'Clean Energy Future' scheme, like emissions trading schemes in the European Union, involves allocating free permits to some polluters. This can lead to charges of political manipulation by well connected industries, which appear to be justified in some instances. Nonetheless, the question remains as to what place, if any, free permits would have in a well executed emissions trading scheme.

There are two potential policy motivations for issuing free permits. First, a fixed quantity of permits may be allocated in order to compensate the owners for reducing the value of a business or asset, such as a power plant. Secondly, a stream of permits may be allocated each year in proportion to the quantity of goods or services produced by a business. This can be a way to ensure that incentives exist to reduce emissions, without affecting the competitiveness of the business relative to other countries that lack carbon regulation. Businesses receiving a stream of permits proportional to their production (for example, a certain number of permits per unit of steel manufactured) nonetheless retain an incentive to reduce their emissions because any emissions savings free up permits that they can sell on the open market.

5 Compensation for loss of asset value

The first motivation - compensation for loss of asset value - applies only to businesses that are not exposed to competition from other countries, such as electricity generators. Owners of some electricity generation assets have argued that they should be compensated for the new costs imposed on their business by carbon regulation. This fairness argument is sometimes accompanied by a practical claim that, without compensation, they may be unable to continue operations, financial markets may be disrupted and the increased perception of sovereign risk may further increase financing and consumer costs into the future. These are said to amount to a risk to energy security unless partial assistance is provided in relation to the most extreme instances of asset value loss.

An important question is whether such permits would continue to be allocated if operations cease or if they would be linked to ongoing production. If the government intends to compensate power generation companies, it is much better to hand over a fixed quantity of permits 'without strings attached' than to link them to ongoing production. This is because linking them to ongoing production sets up a perverse incentive for older, dirtier facilities to continue running so that they can continue to receive valuable, tradable permits.

Preserving the incentive to reduce emissions, including by decommissioning the oldest and dirtiest plants rather than extending their operating lives, requires a pre-determined, one-off allocation of permits, which the owners should be allowed to sell if they choose to shut down their plant. This means that the free permits will, in theory, have no effect on whether the plants continue operating or not – and the primary outcome of the free permits is to transfer wealth to the owners of the assets. The real question then is whether such a wealth transfer is justified. The case in favour is not strong.

The underlying economics of all industries is subject to constant change from many factors – exchange rates, competitive pressures, energy prices and regulatory shifts among them – and these are risks that investors knowingly take on when they make investments, which is why they receive appropriately risk-weighted returns. To compensate businesses for loss of asset values is to recognise that some underlying right exists for them to impose the costs of greenhouse gas pollution on society without limit, and that denying them this right requires compensation.

This would be an unusual right even if the risk posed by greenhouse gases were unheard of at the time the assets were constructed – after all, ignorance has not been accepted as a defence in the case of other harmful externalities which have become known over time, such as asbestos manufacture. While some of Australia's power stations are old (some of the dirtiest were built in the 1980s or earlier), business has been on notice for more than two decades now since the urgency of reducing carbon emissions became apparent, and these assets have been generating profits throughout this time.

In fact, far from compensating polluting and harmful industries when regulations are introduced, the opposite is done in many instances. Businesses are often required to compensate society, even when the degree of harm they imposed was not always clear to them. This is routinely true in cases of direct harm to identifiable individuals, such as the cases of asbestos manufacture, cigarette production, and other forms of harm to workers and consumers.

In addition, companies causing environmental damage are also often held liable for cumulative historic damage, and asked to pay for the cleanup of local waterways and lakes, or to remediate production sites. By not being asked to pay for their historic emissions, power generators have already been treated more generously than other harm-causing businesses, since they profited from the use of a limited common resource (the atmosphere) without paying any direct price for doing so. For the community not only to ignore their historic harm, but to go further and compensate them for regulating their emissions in the future, appears unnecessarily generous to the owners of these emissions-intensive assets.

Other parties appear more deserving of compensation for the economic impacts of carbon pricing. Energy prices are likely to rise as a result of utilities passing on the costs of carbon permits to consumers. Low income households will be disproportionately affected by these rises, since they spend a higher proportion of their incomes than others on utility bills. This means that, without compensation, the cost of reducing emissions may be paid disproportionately by those least able to afford it.[66]

Rebates, compensation or concessions for the poor should therefore be factored into government policy, whether administered through the social security system or directly through utilities. The money raised by the government from selling permits can help to fund the necessary social adjustments. It is crucial, however, that this compensation is structured so that it does not counteract the incentive to reduce usage. In general, this means using a fixed amount of cash compensation, ensuring that the amount of cash transferred does not increase with increased energy usage.

6 Preservation of production incentives and 'carbon leakage'

For industries exposed to competition from other countries, carbon regulation can in principle add costs that may reduce their relative competitiveness. This could potentially lead to reduced production in Australia and increased production overseas. In some cases it might lead to the relocation of production offshore to places where carbon emissions are less regulated. This would cost Australia income and jobs, while not reducing overall global emissions – an undesirable effect known as 'carbon leakage'. Emissions-intensive export industries could be affected, but so, too, would any emissions-intensive domestic industry whose production could be replaced by imports.

Carbon leakage is more than just a theoretical concern. The Spanish concrete industry suffered significant losses following the introduction of carbon regulation there as its domestic production was replaced by imports from neighbouring Morocco. Many of Australia's export industries are already only borderline competitive. This is often attributed to higher labour and environmental costs in Australia, but in reality it is at least as much a result of a lack of economies of scale flowing from limited domestic demand.

Streams of free permits linked to the quantity of goods produced can be used to remove incentives to reduce or relocate production offshore. Linking the number of permits allocated to the quantity of production preserves the incentive to reduce emissions intensity per unit of production, but does not create an incentive to reduce the number of units produced. For example, a steel producer may be allocated one tonne of emissions permits per tonne of steel produced. The actual number can be determined by current best practices in that industry, to set the benchmark at a level that can be realistically achieved. If production shuts down or moves offshore, the allocation of free permits should also cease.

In cases of truly trade-exposed industries, production-linked allocations of permits can be justified because they preserve some environmental incentives while meeting a legitimate economic concern. This justification exists only as long as competing countries do not impose carbon regulations on competing businesses, and such measures must otherwise be applied in a manner consistent with world trade law (see the previous chapter). This is why it is important for such assistance to be regularly reviewed against comparable carbon costs operating in relevant markets and Australia's plan commissions the Productivity Commission to undertake this task.

The problem of carbon leakage is just one good reason why international cooperation on a coordinated global emissions trading scheme is preferable to a series of unilateral national schemes, although the latter is better than nothing. The inter-relationship between any Australian scheme and a global mechanism such as Kyoto's CDM will be important to get right – for example, whether CDM offsets will be able to count towards compliance under the Australian scheme.[67]

I Conclusion

Australia has been slow to act domestically and to show leadership internationally on climate change. While the Australian government supported the negotiation of the UNFCCC framework in the 1980s and early 1990s, a change of government resulted in Australia negotiating special concessions for itself in the Kyoto Protocol and then not even ratifying it. A further change of government led to Australia's ratification of the Kyoto Protocol in 2007 and efforts to design a national emissions trading scheme, which fell apart in 2010 in the face of difficult political conditions.

The integrity of the Australian government's proposed CPRS was seriously eroded as a result of the deep concessions given to special industry groups during a highly politicised policy process. The CPRS came to include price caps and a broad range of free permit handouts, even to businesses such as energy producers which are not threatened by import competition. It exhibited many of the hallmarks of a bad emissions trading scheme. If it had passed, it would have done something to reduce carbon emissions, but it would have needed significant reform over time to be truly effective. It certainly did not meet the tests for environmental and economic effectiveness outlined in the Garnaut Review.

By late 2010, Australia found itself in domestic and international climate change policy limbo, with a domestic emissions trading scheme apparently stalled until at least 2013, and international negotiations after Copenhagen and Cancún proceeding at a painfully slow pace. Faced with a hostile Opposition, and an increasingly disengaged or cynical electorate, political momentum for climate change action in Australia dissipated after a wave of interest in it helped to elect the Rudd government in late 2007.

The failure of Prime Minister Rudd to satisfy the public's demand for action resulted in his declining political fortune and his replacement by Prime Minister Gillard. The new Gillard government shelved immediate action on climate change in the face of an imminent election, but soon after her election victory the Prime Minister established a new process for considering policy options, leading to the February 2011 announcement of a phased scheme to impose a carbon price in 2012 followed by an emissions trading scheme some years later.

As the political debates rage, so does the intensification of extreme weather events continue – think of the Queensland floods in January 2011, the Victorian bushfires of 2009 or the prolonged droughts which brought water restrictions to many cities in the 2000s. Such events press home the reality that climate change continues to affect dramatically the environment on which Australians depend for their livelihoods and their future. The science tells us what is happening; economists and policy-makers give us options for responding; leadership must come from politicians; and the public must encourage them, and trust them, to exercise it.

Notes

1 Rayfuse, R and Scott, S, 2008, 'Australia's Climate Change Law and Diplomacy' 11(1&2) *Asia Pacific Journal of Environmental Law* 1 at 2.

2 Garnaut, R, 2008, *The Garnaut Climate Change Review: Final Report*, Cambridge University Press, Melbourne, p 255.

3 Hamilton, C, 2007, *Scorcher: The Dirty Politics of Climate Change*, Black Inc Agenda, Melbourne, p 74.

4 Pearse, G, 2007, *High & Dry: John Howard, Climate Change and the Selling of Australia's Future*, Penguin Group, Melbourne, p 77.

5 Wilder, M and Miller, M, 'Carbon Trading Markets: Legal Considerations' in Bonyhady, T and Christoff, P (eds), 2007, *Climate Law in Australia*, Federation Press, Sydney, pp 67, 71.

6 Australian Department of Climate Change and Energy Efficiency, 2010, 'Legal Architecture for a Post-2012 Outcome', available at <http://www.climatechange.gov.au/government/international/~/media/submissions/international/Legal-Architecture-Post-2012-Outcome-Australian-submission-MAY-09.pdf>.

7 Senator the Hon Penny Wong, Minister for Climate Change and Water, 2010, 'Australia's Submission to the Copenhagen Accord: Media Release', available at <http://www.climatechange.gov.au/~/media/Files/minister/wong/2010/media-releases/january/mr20100127.ashx>.

8 Australian Department of Climate Change and Energy Efficiency, 2011, 'National Targets', available at <http://www.climatechange.gov.au/government/national-targets.aspx>.

9 Christoff, P, 'Can the Invisible Hand Adjust the Thermostat? Carbon Emissions Trading in Australia' in Bonyhady and Christoff (eds), above n 5, pp 82, 92.

10 Government of New South Wales, 'Greenhouse Gas Reduction Scheme', available at <http://www.greenhousegas.nsw.gov.au>.

11 Essential Services Commission of Victoria, 2011, 'Victorian Energy Efficiency Target', available at <http://www.esc.vic.gov.au/public/VEET>.

12 NSW Independent Pricing and Regulatory Tribunal, 2009, *Compliance and Operation of the NSW Greenhouse Gas Reduction Scheme During 2008: Report to Minister* (July 2009), p 2, available at <http://www.greenhousegas.nsw.gov.au/documents/SchRep08.pdf>.

13 Essential Services Commission of South Australia, 'Residential Energy Efficiency Scheme', available at <http://www.escosa.sa.gov.au/consumer-information/residential-energy-efficiency-scheme.aspx>. In 2009, NSW also implemented an Energy Savings Scheme to assist households and businesses reduce their energy consumption and electricity costs: Government of New South Wales, 'Energy Savings Scheme', available at <http://www.ess.nsw.gov.au>.

14 Macintosh, A, Wilkinson, D and Denniss, R, 'Climate Change' in Aulich, C and Evans, M (eds), 2010, *The Rudd Government: Australian Commonwealth Administration 2007-2010*, ANU E-Press, Canberra, ch 11.

15 Ibid.

16 Department of Climate Change in McKinsey & Company, 2008, *An Australian Cost Curve for Greenhouse Gas Reduction*, McKinsey & Company, Sydney, p 8.

17 Quoted in Macintosh, Wilkinson and Denniss, above n 14, ch 11.

18 Multi-Party Climate Change Committee, 21 December 2010, 'MPCCC Agreed Principles to Guide Development of a Carbon Price Mechanism', available at <http://www.carbontradexchange.com/MPCCC-principles.pdf>.

19 Australian Government, 2011, *Securing a Clean Energy Future: The Australian Government's Climate Change Plan*, p 11.

20 Including a Clean Energy Finance Corporation (with $10 billion for commercial investment), an Australian Renewable Energy Agency (to coordinate $3.2 billion in existing grants) and a Clean Technology Innovation Program (providing grants of up to $200 million to support business investment).

21 Australian Government, above n 19, p 66.

22 Lane, S, 'Treasury Releases Carbon Tax Cost Estimates', ABC News, 1 April 2011, available at <http://www.abc.net.au/news/2011-04-01/treasury-releases-carbon-tax-cost-estimates/2631424>.

23 Abbott, T, Hunt, G and Hockey, J, 10 July 2011, 'Julia Gillard's unnecessary and costly $9 billion tax will hit families: press release', available at <http://www.liberal.org.au/Latest-News/2011/07/10/Gillard-Tax-Hits-Families.aspx>.

24 Truss, W, 10 July 2011, 'Carbon Tax ripple will grow to swamp business & families: press release', available at <http://www.liberal.org.au/Latest-News/2011/07/10/Carbon-tax-ripple-will-grow-to-swamp-business-and-families.aspx>.

25 Ibid.

26 'Everyday cost of carbon price to be minimal, energy bills worst hit', 19 September 2011, *Climate Spectator*, available at <http://www.climatespectator.com.au/news/everyday-cost-carbon-price-be-minimal-energy-bills-worst-hit>.

27 Lane, above n 22.

28 *The World Today*, 11 July 2011, 'Economists question household compensation', *ABC News*, available at <http://www.abc.net.au/worldtoday/content/2011/s3266280.htm>.

29 Abbott, Hunt and Hockey, above n 23.

30 Truss, above n 24.

31 Hunt, G, 11 July 2011, 'Carrots not Sticks: Rethinking Global Approaches to Climate Change: Speech to the Lowy Institute', available at <http://www.liberal.org.au/Latest-News/2011/07/11/Carrots-not-Sticks.aspx>.

32 See, for example, Australian Chamber of Commerce and Industry, 2011, 'Carbon: An Issue About Tax, Not the Environment', available at <http://www.acci.asn.au/getattachment/73026495-9ab0-430a-ac96-040c958009c4/Carbon---An-Issue-about-Tax,-not-the-Environment.aspx>.

33 Ibid.

34 'Carbon Tax will Hit Victoria Hard', 20 September 2011, *Climate Spectator*, available at <http://www.climatespectator.com.au/news/carbon-tax-will-hit-victoria-hard-says-baillieu>.

35 'Calls for halt to carbon tax scaremongering as report predicts minimal impact for Qld', 23 August 2011, *Climate Spectator*, available at <http://www.climatespectator.com.au/news/calls-halt-carbon-tax-scaremongering-report-predicts-minimal-impact-qld>.

36 'Carbon tax to cost NSW billions, jobs: State Treasury', 4 August 2011, *Climate Spectator*, available at <http://www.climatespectator.com.au/news/carbon-tax-cost-nsw-billions-jobs-state-treasury>.

37 'Report Debunks Vic Government's Carbon Tax Impact Claims', 21 September 2011, *Climate Spectator*, available at <http://www.climatespectator.com.au/news/report-debunks-vic-govts-carbon-tax-impact-claims>.

38 Truss, above n 24.

39 See Wood, T and Edis, T, September 2011, *New Protectionism Under Carbon Pricing. Case Studies of LNG, Coal Mining and Steel Sectors*, Grattan Institute, available at <http://www.grattan.edu.au/publications/101_report_new_protectionism.pdf>.

40 Treasury of Australia, 2011, *Strong Growth, Low Pollution*, available at <http://www.treasury.gov.au>.

41 Access Economics, 2009, *Report 2: Impacts on disadvantages regions*, available at <http://www.caf.gov.au>.

42 Treasury of Australia, 14 July 2011, 'Treasury Executive Minute: Economic and Fiscal Impacts of the Coalition Direct Action Plan', available at <http://www.treasury.gov.au/documents/2149/PDF/TEM_coalitions_direct_action_plan.pdf>.

43 Garnaut, above n 2, p 65 (based on the year 2005), other country figures from World Resources Institute.

44 Ibid, p 55 (based on the year 2004).

45 Hunt, above n 31.

46 It claims that the cost to businesses under the EU scheme was a total of $4.9 billion during the first six and a half years, and that the equivalent figure in Australia would be $71 billion: 'Combet slams anti-carbon tax "rubbish", as poll shows voter sentiment shifting', 24 July 2011, *Climate Spectator*, available at <http://www.climatespectator.com.au/news/combet-slams-anti-carbon-tax-rubbish-poll-shows-voter-sentiment-shifting>.

47 Ibid. Free permits reduce the cost to business to $35 billion. The EU in fact will reap a minimum of $143 billion from its scheme in the coming six years.

48 Ross Garnaut, quoted speaking to a federal parliamentary committee inquiry in 'Citizen Garnaut slams carbon critics over "extreme distortion"', 28 September 2011, *The Australian*, available at <http://www.theaustralian.com.au/national-affairs/carbon-plan/citizen-garnaut-slams-carbon-critics-over-extreme-distortion/story-fn99tjf2-1226148483884>.

49 McKinsey & Company, above n 16, p 17.

50 Stern, N, 2006, *Stern Review on the Economics of Climate Change*, UK Government, London, p xii.

51 Ibid, p ix.

52 Garnaut, above n 2, 258.

53 Ibid, p 260.

54 Ibid, p 468.

55 McKinsey & Company, above n 16, p 8.

56 See, eg, ibid.

57 Ibid, p 8.

58 Ibid, p 8.

59 Ibid.

60 Ibid, pp 51–53.

61 Ibid, pp 55–57.

62 Ibid, pp 48–51.

63 McKinsey & Company, 2009, *Pathways to a Low-Carbon Economy: Version 2 of the Global Greenhouse Gas Abatement Cost Curve*, McKinsey & Company, p 29.

64 Australian Parliamentary Library, 19 November 2010, 'Carbon taxes', available at <http://www.aph.gov.au/library/pubs/climatechange/responses/economic/carbontax.htm>.

65 Wilder and Miller, above n 5, p 67.

66 Garnaut, above n 2, ch 16.

67 Wilder and Miller, above n 5, pp 67, 71.

Chapter 5

Climate Change 'Refugees'? Climate-Related Displacement and Migration

Key Issues

- Climate change is unlikely to cause displacement on its own this century, but may exacerbate pre-existing socio-economic stressors which cause people to move.
- Estimates about the numbers of people who may move in relation to climate change impacts are unreliable. This is because of methodological difficulties in determining who to count; the challenge of predicting how populations will adapt at home (instead of migrating); and an absence of bureaucratic categories (eg, visas) for accounting for cross-border and internal climate-related migration.
- Climate-related migration may be triggered by sudden onset disasters (such as extreme weather events) or slower onset processes (such as rising sea-levels and erosion).
- International law does not currently provide specific rules for dealing with people forced to move to another country because of climate-related factors. They do not generally qualify as 'refugees' or 'stateless persons'.
- The planned relocation of entire populations is very contentious and complex. Relocating populations fear losing their culture and traditions, and their situation in the new country will be precarious unless they can secure a permanent immigration status.
- Human rights law sets out minimum standards of treatment for all individuals. It sometimes also provides a basis for claiming protection in another country. However, given the high threshold for claiming such protection, it is unlikely to be of much assistance in the climate change context.

- Most people displaced by climate-related impacts will migrate internally, rather than across an international border. The responsibility for their protection lies with their own governments, which often makes it difficult for other countries and international organisations to render assistance. Where a country is unable or unwilling to assist internally displaced persons, international protection may become necessary.
- It is difficult to hold any country or entity legally responsible for climate change, given the difficulties of establishing whether emissions causing climate change were illegal when emitted; causation (that is, who specifically caused the particular harm done); and whether the specific harm was reasonably foreseeable.
- There is currently no single institution responsible for climate-related migration, although some informal arrangements have been concluded between various international bodies.
- There remains a need to better define legal and institutional frameworks for responding to climate-related displacement at the international, regional and national levels, despite the complexity of this task.

A Introduction

Human movement caused by environmental factors is nothing new. Environmental disasters and gradual environmental degradation have displaced people in the past and will continue to do so in the future. Moving away from such risks is a normal response to change and often a sign of community resilience – even if it does challenge ideas about border protection. What is 'new' about climate-related displacement is the role of humans in causing climate change,[1] the large number of people affected by it,[2] and the relative speed of it, all of which are likely to hamper people's ability to adapt to changing conditions.

The impacts of climate change are indiscriminate and affect people everywhere, but some parts of the world will feel the effects more acutely than others. Not only are some physical environments more vulnerable to the changing climate, but some governments and communities lack sufficient resources to adapt. Shoreline erosion, coastal flooding, desertification, salinity, the vulnerability of small islands to rising sea-levels and increased severe weather[3] compromise the continued habitability of some places. As shown in Chapters 1 and 2, these effects impact on agricultural viability, vital infrastructure and services, the stability of governance and, ultimately, human settlement.[4] Climate-induced displacement is particularly relevant to Australia given its geographical proximity to islands at risk of rising sea-levels in the South Pacific. Some island nations, such as Kiribati and Tuvalu, are already experiencing adverse impacts and people fear that they may eventually have to leave their homes, with Australia an obvious destination.

The term 'climate change refugee' is sometimes used to describe people who will be forced to leave their homes as a result of climate change. This chapter explains why the use of this term is misleading, but also why there remain real gaps in existing legal frameworks for protecting such people. It considers: how many people are likely to move because of climate change; the role climate change plays in decisions to move; how law and policy should respond; and what institutional changes might be made at the international level to assist displaced people. As will be seen, there is a lively debate about how to deal with the problem. Some advocates call for a new international treaty to protect the displaced, while others focus on a more nuanced mix of responses such as adaptation, migration and assistance.

B The numbers of displaced people

There has been a lot of debate about how many people are likely to be displaced by climate change. Estimates of the numbers are controversial.[5] The first problem is working out who should be counted. There is no agreed definition of a 'climate migrant' or 'climate refugee', and no visa category in any country which classifies people in this way. Debates about the extent to which poverty, limited natural resources, overcrowding and political conflict interact with environmental stressors to force movement necessarily affect who is classified as a 'climate migrant' – as opposed to an 'economic' or 'environmental' migrant, for example.[6]

In addition, there are problems in identifying: (a) the pace at which such change may force people to move; (b) the degree to which movement will occur within a country or across international borders; (c) whether movement will be permanent or temporary; (d) uncertainty about how far carbon emissions will be reduced; and (e) how different communities will adapt to climate change impacts. People who move internally, for instance, are unlikely to be counted as 'climate migrants', because in most countries citizens and residents enjoy freedom of movement and the reasons for their movement are not officially recorded. Likewise, people who move to another country often do so within regular labour, education or family migration programs, so the problem of climate-related migration remains formally invisible to governments and legal systems.

These difficulties, combined with the different methods used by researchers, have led to very different conclusions about the size of the problem. The most commonly cited figures come from the work of a social scientist, Norman Myers. In 1993, Myers suggested that 50 million people could be displaced by climate change by the middle of this century.[7] His methodology was fairly crude: he identified parts of the world thought to be vulnerable to sea-level rise and an increase in extreme weather events and calculated the number of people likely to be affected by them based on anticipated population growth. He did not take into account any adaptation measures.

In 2005, he revised his estimate upwards to 200 million. In 2007, he went even higher – to 250 million.[8] An NGO, Christian Aid, then suggested that one billion people would be displaced by climate change by 2050.[9] This massive figure continues to be commonly cited even though it is not grounded in research, and in fact includes other types of displacement as well. For instance, Sir Nicholas Stern, in his authoritative economic review of climate change in 2007, described Myers' estimates of 200 million as 'conservative'.[10]

Global policy-makers have also suggested that climate change may lead to large-scale forced migration. In 2009, a report by former United Nations Secretary-General Kofi Annan described 'millions of people' being 'uprooted or permanently on the move as a result' of climate change, with '[m]any more millions' to follow.[11]

The United Nations High Commissioner for Refugees, António Guterres, also pointed to the environmental impacts of climate change as the key driver of a surge in future displacement,[12] with the most affected areas being Africa, the mega deltas of Asia, and small island countries.[13]

A recent study of 23 countries and regions by researchers at the United Nations University in Bonn provides a more nuanced and research-based approach. It is important for two reasons. First, it highlights the different types of weather events that may be exacerbated by climate change (such as flooding, riverbank erosion, cyclones and so on). Secondly, it supports the view that the complexity of migration decisions, and the inter-connectedness of environmental, economic, social and political factors, make it virtually impossible to provide an accurate estimate of people displaced 'by' climate change. Importantly, it found that most movement is likely to be internal, rather than across borders,[14] and that longer term, permanent movement is becoming more common than temporary or seasonal relocation, especially among younger people.[15]

It is unknown how many people have already moved in response to climate change impacts, although people certainly have moved already and will continue to do so.[16] Scientific projections about areas at risk of coastal flooding, drought, rising sea-levels and extreme weather events (such as hurricanes, typhoons and king tides) give a good indication of places at risk. For example, in 2008, some 36 million people were displaced by natural disasters, 20 million of whom arguably moved because of 'climate change-related factors'.[17]

C The nature and causes of displacement

1 Climate change impacts driving displacement

Not all climate change related displacement is the same. Broadly speaking, it can result from two kinds of occurrences: sudden onset disasters, such as extreme weather events, and slower onset processes, such as rising sea-levels, erosion and salinity. In the case of sudden events, movement is typically temporary, within the country, and often not very far from home (such as after Hurricane

Katrina in the United States and Cyclone Nargis in Burma). By contrast, slow onset processes occur over a much longer period and may ultimately necessitate permanent movement (in some cases across international borders).

The United Nations Secretary-General's former Representative on the Human Rights of Internally Displaced Persons, Walter Kälin, developed a framework setting out the diverse scenarios that can be encompassed within the wide concept of environmental displacement. Although not all scholars necessarily agree with its categorisations, the framework was adopted by a key United Nations body working on the issue.[18] Each type of scenario described below involves different kinds of pressures and impacts, which will affect the time, speed and size of movement. This means that, at various points in time, the role of climate change in individual or household decisions to move may be stronger or weaker. Since such a wide range of scenarios can be caught under the 'climate displacement' umbrella, no single legal or policy response can adequately address them all.

(a) The increase of hydro-meteorological disasters

This includes flooding, hurricanes, typhoons, cyclones and mudslides, and is likely to lead predominantly to internal displacement. An example is temporary internal migration in response to flooding in the Mekong Delta region in Vietnam.[19]

(b) Government-initiated planned evacuation of areas at high risk of disasters

This is likely to lead to permanent internal displacement. For example, in Mozambique the government has organised 'resettlement programs' to remove people from the physical danger of extreme floods. However, these can create other environmental, social and economic problems, and cause people to lose their livelihoods and become dependent on the government and international aid.[20] It can also cause social conflict if the affected communities are not adequately consulted about the process.

(c) *Environmental degradation and slow onset disasters*

This includes water availability, desertification, recurrent flooding and increased salinity in coastal zones. Kälin explains:

> Such deterioration may not necessarily cause forced displacement strictly defined, but instead incite people to move to regions with better income opportunities and living conditions before it becomes impossible to stay at home. However, if areas become uninhabitable because of complete desertification or sinking coastal zones, then population movements would amount to forced displacement and become permanent.[21]

For example, some argue that permanent relocation is now the only durable solution for certain Alaskan indigenous communities (the Inuit),[22] but there are obstacles to their movement, including: (a) a lack of a government agency with authority to relocate communities or address the strategic planning needs of relocation; (b) a lack of funding designated for relocation; and (c) a lack of criteria for choosing relocation sites.[23] Likewise, in Bangladesh, low-lying delta and river areas are highly vulnerable to sea-level rise, flooding and salinity. A 45 centimetre sea-level rise in Bangladesh could displace 5.5 million people and inundate 11 per cent of Bangladesh's territory, while a 1.5 metre sea-level rise could affect 17 million people (15 per cent of the population) and 16 per cent of the total land area.[24] The stress of displacement would exacerbate other pressures, such as overpopulation, poverty and unequal land distribution.

(d) *Small island countries at risk of 'disappearing' because of rising seas*

If a territory becomes uninhabitable (for example, because of the inability to grow crops or obtain fresh water), permanent relocation to other countries will be necessary even if the country is not yet under water. However, international law does not provide a protected status for such people. Although small island countries (such as Kiribati and Tuvalu) emit less than 1 per cent of global greenhouse gases, their small physical size, exposure to natural disasters and climate extremes, very open economies and low adaptive capacity make them particularly susceptible, and less resilient, to climate change.[25]

The submergence of small island countries is often publicly cited as a clear-cut case of 'climate change' displacement, particularly in the 'sinking' South Pacific islands of Kiribati and Tuvalu which average less than two metres above sea-level. We need to be cautious, though, because the empirical evidence does not support such a straightforward explanation of movement. Rather, climate change may provide a 'tipping point' as its impacts are felt *in combination with* other pressures in these countries. Overcrowding, unemployment, scarcity of fresh water, inadequate infrastructure, pollution and so forth are already compromising the viability of the populations in these countries. One government official in Kiribati likened the effect of climate change on his country to the effect of HIV/AIDS on a person: you don't die *from* HIV/AIDS, but having it makes life much more precarious and greatly increases susceptibility to other illnesses.[26]

(e) Risk of conflict over essential resources

Even though the humanitarian community is used to dealing with internal conflicts, and people displaced by conflict may be eligible for protection as refugees or assistance as internally displaced persons (IDPs) (see further below), resource-based conflicts 'may be particularly challenging'. In particular, where resource scarcity (such as a lack of fresh water or secure, arable land) cannot be resolved, 'it will be extremely difficult to reach peace agreements providing for an equitable solution. The likely outcome is both conflict and the displacement of a protracted nature'.[27] Conflict is more likely to be social conflict than armed conflict. Chapter 6 of this book looks in more detail at the conflict and security risks associated with climate change.

2 Climate change interacts with other pressures

The framework set out above by Kälin is a useful starting point for thinking about how the effects of climate change might stimulate human movement. The reality of movement is complicated and a range of factors influences people's decisions to leave their homes. UNHCR notes that it is increasingly difficult to categorise displaced people because of the combined impacts of conflict, the environment and economic pressures.[28] Certainly, climate change is having real impacts on people's lives, but in most cases it is only one of a number of reasons why people decide to move. Climate change tends to multiply pre-existing stressors, rather than solely causing movement.

In the case of a sudden disaster, people generally flee for their lives. They often do not travel very far or stay there for very long. Whether or not people stay or go will be influenced by their country's economic, social and political capacity to assist them. This is true both in terms of how severely the disaster hits (which is linked to the effectiveness of adaptation or prevention strategies), as well as how lasting its effects are (in terms of the ability to rebuild).

For example, while many internal migrants from north-west Ghana cite environmental reasons for moving, including 'scarcity of fertile land, unreliable rainfall, low crop yields and/or food security problems',[29] they also cite other factors, such as 'lack of non-farm income opportunities, family conflicts, witchcraft, cattle theft and the desire to be free and independent'.[30] Researchers in one study concluded that 'political and economic forces seem to have more influence than environmental push on migration flows',[31] and that '[t]he environmental driver of migration ... appears to be structural scarcity rather than degradation'.[32]

That said, people affected by natural disasters in Bangladesh have often been able to return safely to their homes because of effective early warning systems and storm shelter infrastructure.[33] In the 2011 floods and cyclone in Queensland, relatively few lives were lost because of forewarning, preparation and emergency assistance. Given the severity of those disasters, the human toll would have been far worse had these measures not been in place.

People are less likely to move outside their own country where information and resources are available to help remedy their plight (such as drought or flood relief packages, housing assistance, employment opportunities, social security and so on). This is why Australian farmers, or beachfront residents of Byron Bay whose houses are falling into the sea, are never likely to be seriously discussed in international climate migration debates.[34]

People who want to leave their homes typically only do so where they have the financial means and access to networks that support migration. For example, in Ethiopia, drought is the main environmental stressor, impacting on land productivity. But individual and household decisions about whether or not to migrate are linked to many other issues. Political, structural and other factors, such as ethnicity, family size, productivity in potential destination zones, the agrarian state of rural areas, schooling opportunities and the availability of water elsewhere, all impact on decision-making about migration possibilities.[35]

In effect, individuals 'try to calculate the *relative advantages* of moving against the relative advantages of remaining behind', including 'weighing up the chances of finding work in town against the possibility that one might come across a good deal on renting land in the rural areas ... having friends or family already resident in the urban areas or having an ageing family member with good land in the rural areas'.[36] International movement may even be considered relatively undesirable, because it severs a person's connections with their land, family, culture and community, whether in small Pacific island countries or rural Bangladesh. A person's skills, education and form of livelihood may also not be well recognised or valued elsewhere.

Some researchers therefore argue that it is arbitrary to identify 'climate change' as a driver of forced migration, while omitting other – often more pressing – causes such as poverty, conflict or lack of opportunity.[37] They instead prefer the term 'survival migration', focusing attention away from single causes of movement and towards a more comprehensive understanding of multiple stressors. Indeed, much of the 'responsibility' for movement resulting from things like poverty may also be attributable to certain international structures (economic and political) that perpetuate an ever-growing divide between rich and poor countries.

To summarise then, displacement can result from a complex combination of inter-relating factors, as the table (*over page*) indicates:

Sudden Onset Disasters	Slow Onset Processes	Economic, Social and Political Factors
• Flooding • Cyclones, hurricanes and typhoons • Landslides and mudslides • Fires	• Rising sea-levels • Coastal erosion • Riverbank erosion • Salinity • Droughts • Desertification	• Environmental mismanagement and pollution • Weak governance and poor planning • Resource scarcity • Poverty and unemployment • Failed adaptation • Personal, family and community factors

D The law

While movement in response to environmental and climate change is a normal human adaptation strategy, the difficulty today is that people cannot simply migrate as and when they choose. National immigration laws firmly restrict the entry of non-citizens into other countries. Under international law, too, countries are normally only required to admit their own citizens. International law only recognises a very small class of forced migrants as people whom other countries have an obligation to protect: 'refugees', 'stateless persons', and those eligible for 'complementary protection' (based on human rights law), discussed below. This means that unless people fall within one of those groups, or can lawfully migrate for reasons such as employment, family or education, then they run the risk of interdiction, expulsion and detention if they attempt to cross an international border and have no legal entitlement to stay in that other country.

1 Are climate-displaced people 'refugees'?

The term 'refugee' is a legal term of art. The legal definition of a 'refugee', and the rights and entitlements which a refugee is owed, are set out in the 1951 Refugee Convention relating to the Status of Refugees and its 1967 Protocol. A 'refugee' is someone who:

> owing to well-founded fear of being persecuted for reasons of race, religion, nationality, membership of a particular social group or political opinion, is outside the country of his nationality and is unable, or owing to such fear, is unwilling to avail himself of the protection of that country; or who, not having a nationality and being outside the country of his former habitual residence as a result of such events, is unable or, owing to such fear, is unwilling to return to it.[38]

In other words, a refugee is someone who is already outside his or her country of origin, whose government is unable or unwilling to protect him or her, and who has a well-founded fear of persecution *for reasons of* his or her race, religion, nationality, political opinion or membership of a particular social group. There are four elements of this definition that present obstacles in trying to argue that people at risk of climate change are refugees.

First, the refugee definition only applies to people who have already crossed an international border. As noted above, much of the anticipated movement in response to climate change will be internal and thus will not meet this preliminary requirement.

Secondly, there are difficulties portraying 'climate change' as 'persecution'. 'Persecution' entails violations of human rights that are sufficiently serious, either because of their inherent nature or because of their repetition (for example, an accumulation of breaches which, individually, would not be so serious but which together constitute a severe violation). It remains very much a question of degree and circumstance. Whether something amounts to 'persecution' is assessed according to the nature of the right at risk, the nature and severity of its impairment, and the likelihood of the impairment actually eventuating in the individual case.[39]

Although adverse climate impacts such as rising sea-levels, salination, and increases in the frequency and severity of extreme weather events are harmful, they do not meet the threshold of 'persecution' as this is currently understood in law. Part of the problem is in identifying a 'persecutor'. For example, the governments of Kiribati and Tuvalu are not responsible for climate change as a whole, nor are they developing policies which increase its negative impacts on particular sectors of the population. One might argue that the 'persecutor' in such a case is the 'international community', and industrialised countries in particular, whose failure to cut greenhouse gas emissions has led to the predicament now being faced.

Yet, these industrialised countries are the very ones to which people are likely to move if they can no longer remain at home. This is a complete reversal of the traditional refugee paradigm. Whereas refugees flee from their own government (or private actors that the government is unable or unwilling to protect them from), people fleeing the effects of climate change are not escaping from their government, but are seeking refuge from – but within – countries that have contributed to climate change.

Finally, even if the impacts of climate change could be characterised as 'persecution', the Refugee Convention requires such persecution to be *for reasons of* an individual's race, religion, nationality, political opinion or membership of a particular social group. Persecution alone is not enough. The difficulty here is that the impacts of climate change are largely indiscriminate, rather than tied to characteristics such as a person's background or beliefs. Although climate change more adversely affects some countries, by virtue of their geography and resources, it does not do so due to the nationality or race of the population.

An argument that people affected by the impacts of climate change could constitute a 'particular social group' under the Convention definition would be difficult to establish, because the law requires that the group must be connected by a fundamental, immutable characteristic other than the risk of persecution itself.[40] So far, there have been a small number of cases in Australia and New Zealand where people from Tuvalu and Kiribati have argued that they should receive refugee protection from climate change impacts,[41] but all have failed.

Nevertheless, there remain limited exceptions where exposure to climate impacts or environmental degradation might amount to persecution for a Convention reason. One example would be where government policies discriminatorily target particular groups reliant on agriculture for survival (for example, as a way of intimidating minority or resistance groups), where climate change is already hampering their subsistence. Another example would be if a government induced famine by destroying crops or poisoning water, or contributed to environmental destruction by polluting the land and/or water.[42] However, in most cases people displaced by climate change are unlikely to gain protection as refugees.

2 Will climate-displaced people be 'stateless'?[43]

As mentioned earlier, there has been considerable media attention given to the so-called 'disappearing states' or 'sinking islands' – whole countries that could be submerged by rising sea-levels. The science on rising sea-levels was discussed in Chapter 1. Would people who moved away from such island countries be recognised as 'stateless persons' under international law (provided they had not acquired any other nationality, for example by moving to and becoming citizens of another country)?

In international law, a 'state' exists if a defined territory has a permanent population, an effective government and the capacity to enter into relations with other countries. The 'disappearing islands' idea assumes that the loss of territory – through submergence of land – will signal the end of the country. As noted earlier, however, small island countries are likely to become uninhabitable long before they physically disappear. This means that the absence of population, and with it the loss of effective government, are likely to be the first signs that a country has started to 'disappear' as a legal entity.

However, international law rules on the extinction of countries have never been tested in this way before. International law contemplates the formal dissolution of a country in cases of *absorption* (by another country, such as through a referendum), *merger* (with another country, as in the reunification of West and East Germany) and *dissolution* (with the emergence of successor countries, as in the break up of the former Soviet Union and Yugoslavia). It has never had to deal with the potential extinction of a country because of physical or territorial disappearance.

The reason why this is relevant is because a country's continued existence is linked with its citizens' ability to maintain, and effectively exercise, their nationality. When a country ceases to exist, so does nationality of that country.[44] Importantly, nationality carries with it a bundle of rights and legal protections for citizens.

The legal definition of a 'stateless person' in Article 1 of the 1954 Convention relating to the Status of Stateless Persons is deliberately restricted to individuals who are 'not considered as a national by any State under the operation of its law'.[45] In other words, this relates to situations where a country that has actually deprived someone of nationality. The treaty does not extend to situations of *de facto* statelessness – where a person formally has a nationality but it is ineffective in practice (although the treaty's drafters recommended that it be applied to them).[46]

As the 1954 statelessness treaty stands, it would not protect people whose country is at risk of disappearing, unless the country formally withdrew nationality from them (which, as a matter of human rights law, it is obliged not to do). But if a country no longer exists, then its former population would fall within the 'stateless person' definition, provided they had not acquired a new nationality. This would oblige signatory countries to provide to such people within their territory the rights in that treaty, including 'as far as possible facilitat[ing] the[ir] assimilation and naturalization'.[47]

The practical benefits of this remedy are, however, limited. First, it remains unclear in international law when countries would be prepared to regard an established state as having 'disappeared'. This is because history shows that the international community tends to presume the continuity of existing countries, even when some of the formal criteria of statehood start to wane. Secondly, relying on the statelessness treaty is reactive rather than proactive, because it is only 'triggered' once a person is physically present in another country's territory. This means that people have to leave their homes and get to another country before being able to claim its benefits. Finally, the treaty is very poorly ratified and most countries do not have any formal procedures for determining the legal status of stateless persons. Thus, there is no clear means by which the treaty's benefits could be accessed.

However, the advantage of the statelessness concept is that UNHCR has a mandate to prevent and reduce statelessness.[48] This means that an international organisation is empowered to advocate on behalf of affected populations and talk to states about preventing statelessness and assisting stateless people.[49] By contrast, it has no mandate with respect to people displaced by climate change in other contexts, since they are not 'refugees'.

3 Should people be relocated *en masse*?

The relocation of a country's entire population has sometimes been suggested as a response to potential whole-nation displacement. This raises a number of significant issues relating to self-determination (the right of a population freely to choose its political status, including independence as a separate country), cultural identity and nationality.

Interestingly, the idea of whole country relocation is not really supported by the governments of Kiribati and Tuvalu (two countries often mentioned in this context). This is mainly because they would like to remain in their own countries, and also because they realise the political unlikelihood of relocation ever happening. Previous relocations in the Pacific (and elsewhere) have rarely been successful. How to balance the needs of those who move with the needs of the community they move into is very difficult, especially in the Pacific islands where spiritual ties to land mean that land cannot simply be bought and sold to anyone.[50]

More generally, though, there is more to relocation than just securing a place to live. People who move need to know that they are able to remain in the country, work, have access to healthcare and social security, and maintain their culture and traditions. In other words, they need immigration or citizenship rights (including for children born there). When Australia offered to relocate the population of Nauru in the 1960s and grant them Australian citizenship (but not their own sovereignty), the offer was rejected.[51] Concerns about maintaining identity, culture, social practices and land tenure are very real to those whose movement is proposed, and these may not be readily understood by outsiders. This, in turn, may lead to misunderstandings and misguided policies.

It is only with formal cession of land (that is, handing over sovereign legal ownership of land) at the country-to-country level that a country can acquire the lawful international title to it and thus enable its nationals to move to that area as part of their own national territory. The likelihood of this happening today is remote.

It is hard to imagine, for example, Australia yielding sovereignty over any of its own territory or islands so that displaced Pacific Islanders could recreate their own state there.

Surprisingly, in the aftermath of the 2010 earthquake in Haiti, the President of Senegal offered to resettle displaced Haitians as descendants of African slaves: 'They have a right to return to Africa, their original land. They were colonized by the Americans. We will find them land'.[52] It was also suggested that Haitians seeking a new nationality should be naturalised.[53] The African Union is purportedly considering a proposal to create a new country for them in Africa, citing 'a sense of duty and memory and solidarity'.[54] In this context, historical ties and symbolism of 'home' are paramount, even if distant. They are not present in other contexts where relocation has been mooted, such as between Australia and Kiribati or Tuvalu.

However, even where historical connections exist, relocation plans have rarely been successful. Sierra Leone, established in 1787 by Britain as a homeland for former slaves, and Liberia, created for freed American slaves in 1821, both saw unrest and clashes between the indigenous population and the newcomers, which can be traced through to very recent conflicts.[55] All of this suggests the need to be cautious to ensure that any relocation proposals also fundamentally protect the identity, rights and freedoms of the displaced.

4 How does human rights law apply?

International human rights law is important to climate-induced displacement for three reasons. First, it sets out minimum standards of treatment that countries must afford to individuals within their territory. Secondly, if those rights are at risk, human rights law may provide a basis on which individuals can claim protection in another country. This is based on the expanded principle of *non-refoulement* (non-return) under human rights law, embodied in the concept of 'complementary protection' discussed further below. Thirdly, if people move, human rights law requires minimum standards of treatment to be observed in the host country (including for non-citizens) and it is thus relevant to the legal status of the displaced. While at a bare minimum they must not be subjected to inhuman or degrading treatment, countries must also guarantee them the full range of human rights.

There is no express human right to a clean or healthy environment, although a number of human rights necessarily require an environment of a certain quality in order to be fulfilled.[56] As Judge Weeramantry stated in the International Court of Justice:

> The protection of the environment is … a vital part of contemporary human rights doctrine, for it is [an indispensable requirement] … for numerous human rights such as the right to health and the right to life itself.[57]

A number of reports, including by the UNHCR in 2009[58] and the Australian Human Rights Commission in 2008,[59] have examined the impact of climate change on human rights. For example, extreme weather events can impact on the right to life; increased food insecurity can threaten the right to adequate food; water stress can jeopardise the right to safe drinking water; the adverse health effects of climate change can impair the right to health; and sea-level rise and flooding can harm the right to adequate housing.

The human rights impacts of climate change were well illustrated in a petition lodged by the Inuit indigenous peoples of the Arctic regions of the United States and Canada, before the Inter-American Commission on Human Rights (a regional human rights protection mechanism). They argued that the effects of climate change on their environment, allegedly caused by the acts and omissions of the United States (in particular, its failure to ratify the Kyoto Protocol), violated basic rights, including the rights to life, health, physical integrity, security, a means of subsistence, property, inviolability of the home and enjoyment of the benefits of culture.[60]

The Inuit explained that animals on which they relied for food were disappearing, damaging their subsistence harvest and health. Thawing permafrost was causing landslides and complicating food storage. Travel was becoming increasingly dangerous and difficult due to unpredictable weather, with the warmer climate making traditional knowledge about the safety of sea ice unreliable (leading to an increased number of drownings each year). Though the Commission dismissed the petition, thus avoiding contentious litigation against the United States, it did permit the Inuits the opportunity to outline their claims at a hearing and to explain how environmental harms can impact on human rights.

Indigenous peoples may face particular forms of harm. Where people are forced to move away from their traditional lands, the rights of ethnic, religious, linguistic or indigenous minorities to enjoy their own culture, practise their own religion and use their own language may be jeopardised. The Inter-American Commission on Human Rights has acknowledged that 'the use and enjoyment of the land and its resources are integral components of the physical and cultural survival of the indigenous communities',[61] and that interference with these cultural rights may lead to forced assimilation.[62]

(a) Protection from human rights violations?

Could exposure to human rights violations, such as those outlined above, give rise to a claim to legal protection in another country? Certainly, human rights law has expanded countries' protection obligations beyond the 'refugee' category, to include people at risk of arbitrary deprivation of life, torture, or cruel, inhuman or degrading treatment in their home countries. This is known in international law as 'complementary protection' (or 'subsidiary protection' in some countries), because it describes human rights-based protection that is complementary (or additional) to that under the 1951 Refugee Convention.

Only with the passage of the *Migration Amendment (Complementary Protection) Act 2011* (Cth) in September 2011 are non-citizens in Australia permitted to claim protection on these grounds. A person in need of complementary protection can now be granted a protection visa (which is the same visa a Convention refugee receives) to remain permanently in Australia and to build a new life there.

However, it is not clear that a person whose human rights are jeopardised by climate impacts specifically would meet the criteria for such protection. Like the refugee definition discussed above, there is an assumption that the harm feared will be suffered at the hands of a human being rather than the forces of nature (sometimes illustrated by an 'intent to cause harm' requirement). It is doubtful whether violations which are not inflicted by the home country will be seen as giving rise to protection.

Courts have carefully limited the meaning of 'inhuman or degrading treatment' so that it cannot be used as a remedy for general poverty, unemployment, or lack of resources or medical care, except in the most exceptional circumstances.[63] Although the existing law does not *preclude* the impacts of climate change from being recognised as a source of inhuman treatment (for example), it would need to be substantially developed before such harms would fall clearly within the scope of this concept.[64]

Further, for policy reasons (such as to prevent the 'floodgates' from opening), some domestic complementary protection schemes 'carve out' exceptions, which mean people are denied a visa to remain. Such exceptions may include, as in Australian law, situations where the risk is faced generally by a whole population. This raises important global policy questions concerning the equitable sharing of the burden of responding to mass displacement. Without international cooperation, there is a risk that countries close to the source of displacement will have to bear disproportionately the costs of assisting or resettling the displaced, while more distant countries will avoid such costs.

5 What about IDPs?

It is generally accepted that the majority of climate-related movements will be internal. On the one hand, this means that legal responses are simpler, in that we do not need new international treaties to determine responsibility for the displaced. Countries have human rights obligations to all displaced people within their territory, and most people will also have rights as citizens or residents as well.

On the other hand, this makes it difficult for other countries or organisations to 'reach' those affected, since they cannot intervene or operate in another country without its permission. Whereas people who cross an international frontier may engage the protection or assistance of other countries and/or international organisations, IDPs do not enter the international sphere of protection in the same way.

For people who move within their own country, the 1998 United Nations Guiding Principles on Internal Displacement are relevant. Although this document is not itself legally binding, it draws on binding principles from international refugee law, human rights law and humanitarian law, and applies them specifically to internally displaced persons. The Guiding Principles help countries understand what they need to provide to IDPs to assist and protect them at all stages of displacement – from preventing displacement in the first place, to addressing needs during displacement, and in the return and recovery phase.

The Guiding Principles expressly encompass people who have fled their homes due to natural or man-made disasters, although UNHCR's institutional mandate as the lead agency responsible for IDPs only applies to IDPs displaced *by conflict*. The International Committee of the Red Cross, too, is normally only concerned with those displaced by conflict, although national Red Cross and Red Crescent Societies assist with natural disasters.

In 2009, the African Union (a regional organisation of African countries) innovatively adopted the Kampala Convention on Internally Displaced Persons,[65] which is not yet in force. This treaty requires signatory countries to 'take measures to protect and assist persons who have been internally displaced due to natural or human made disasters, including climate change'.[66] While this is the first treaty expressly to recognise climate change as a form of 'natural or human made' disaster, it essentially just elaborates on the IDP definition in the United Nations Guiding Principles (which refers to 'natural or human-made disasters'), and only deals with internal displacement not international movements.

Where a country is unwilling or unable to assist its own people displaced by climate change (for instance, because it lacks resources or discriminates against minority groups), the need for international protection or assistance becomes acute. Yet there is no right in international law for one country forcibly to intervene in another country for humanitarian reasons. It then becomes all the more important that the United Nations Security Council takes action, such as by pressuring a country to accept outside assistance or otherwise authorising appropriate intervention (in very severe cases) to assist vulnerable people in great peril. The role of international institutions in responding to climate-related displacement is considered below. Before turning to that issue, one final legal issue deserves mention – whether litigation in the courts can assist those displaced by climate-related factors.

6 Can anyone be held legally responsible for 'causing' displacement?

So far, the legal analysis in this chapter has focused on the immediate question of how the law can protect or assist those displaced by climate-related factors. An underlying question is whether the law can identify any legal responsibility of particular countries or other entities (such as corporations) for 'causing' climate change related harm, and therefore attribute liability to pay compensation or provide other remedies to the displaced.

In legal terms, it is very difficult to establish causation for climate change. While climate science indicates that there is no plausible explanation for certain weather patterns other than human-induced changes to the climate – the law typically demands a more precise test. It is not enough simply to show that 'industrialised countries' have largely caused climate change through their historical and current carbon emissions.

The foremost problem is that litigation normally requires the claimant to establish that the alleged conduct was illegal or otherwise wrongful at the time it was performed. This immediately presents difficulties, because the law of most countries, and international law itself, did not contain any rules prohibiting carbon emissions until very recently. Consequently, it is very difficult to establish legal liability for historical emissions (say, before the United Nations Framework Convention on Climate Change (UNFCCC) or Kyoto Protocol in the 1990s) which have predominantly contributed to climate change, because such emissions by companies or governments at that time were not unlawful. The case for liability only becomes stronger following the adoption of emissions reductions targets since the 1990s, from which point excessive emissions might well entail legal liability.

Even if an underlying wrong can be identified, however, there are further legal difficulties in suing. The narrow focus of most litigation is not really an effective way of addressing the issue. Ordinarily claimants have to show that particular acts by particular actors at particular times have led to particular consequences for particular groups. For example, using a domestic law framework of negligence law, experts have suggested that anyone seeking to sue for damage arising from climate change 'would face formidable legal, logistical, evidentiary and financial obstacles'.[67]

First, there are significant difficulties in establishing that the resulting harm was a reasonably foreseeable consequence of a particular defendant's activities. The global nature of climate change, and the fact that alleged harms may be 'geographically and temporally divorced from the adverse consequences', means that it is difficult for a claimant to show a sufficiently direct or specific relationship to establish a duty of care.[68] To put it starkly, does a polluting power station operator in Australia really owe a duty of care to indigenous peoples in Alaska, in the event that emissions here affect livelihoods there? The law does not usually stretch so far, even if morally one thinks it ought to. There is a question whether courts would be prepared to recognise such a duty as a matter of legal policy, given its potentially far-reaching consequences. Doing so would amount to a 'revolutionary rather than an evolutionary development of the law'.[69]

Thirdly, even if a duty were recognised, it could be very difficult to establish that it had been breached. This is because a court or tribunal would have to decide whether the particular harm was reasonably foreseeable – should a reasonable person in the defendant's position have anticipated such a risk, particularly when the scientific risks of carbon emissions have only become well known fairly recently? At the same time, since knowledge about the impacts of climate change has developed over time, it is arguable that the requisite standard of care should increase over time as well.[70]

Fourthly, establishing causation is likely to 'pose the greatest obstacle'[71] because of the difficulty in showing an evidentiary link between the negligence and the harm, and attributing responsibility according to legal principles. Eliciting expert scientific evidence to prove this is likely to be fraught with difficulty. This is because 'the impacts involve intensification of existing climatic phenomena, such as more frequent storms, rather than the creation of unique "signature diseases" such as asbestosis, which is caused only by exposure to asbestos'. Further, 'because climate is affected by several factors interacting in complex ways, it is difficult for scientists to tease out what percentage of any climate change is affected by greenhouse gases, and it is even more difficult to determine what percentage is affected by a specific polluter or group of polluters'.[72]

Similar problems arise when approaching the issue from a wider international law perspective.[73] It is not only necessary to establish that a particular state is responsible for harm and to disentangle it from the responsibility of other emitting states (including as regards historical and contemporary emissions), but also to identify what specific legal rule has been violated.

Further, certain areas of international law contain other limitations. Under international human rights and refugee law, for example, countries generally only owe obligations to people already in their territory or jurisdiction. This means that while the United States, for instance, might be considered responsible for its emissions which breach the human rights of people within its territory, it would be far more difficult to demonstrate that these same emissions, drifting ethereally and randomly as carbon particles around the world, constitute violations of the rights of distant populations (see the Inuit claim above).

In addition to all of these factors, foreign countries typically enjoy immunity from being sued before the domestic courts of other countries, so it will usually be very difficult to find a forum in which a case could even be brought. At the international level, the jurisdiction of the International Court of Justice is quite limited, and it is difficult to find any other relevant international courts or tribunals (perhaps apart from certain human rights bodies) that would have the legal authority to hear such cases.

To summarise, then, the table below sets out the key legal categories – and the limits of those categories – in relation to people displaced by climate change related factors:

Legal category	Protection	Limitations
Refugee	• Protects a person from return to a place of persecution	• Persecution is unlikely to encompass victims of carbon pollution, unless discriminatory
Complementary protection	• Protects a person from arbitrary deprivation of life or cruel, inhuman or degrading treatment	• It is unclear whether climate-related harms would be recognised as such
IDP	• Rights-based framework to protect all persons displaced within a country	• Guiding Principles are not binding and cannot be enforced against a neglectful government
Litigation	• Potentially offers a binding judicial decision with compensation or other remedy	• Foreign governments cannot be sued in local courts; international courts may not have jurisdiction

E International responses

1 Institutional governance

Institutional interest in the links between climate change and migration has increased considerably in the past few years. Though there is an absence of coordinated global policy, the connections between climate change and human movement have been acknowledged by international and regional organisations, and in some cases expressly referred to in resolutions and declarations of key bodies. Some of these are discussed below.

There is at present no authoritative or centralised international institution responsible for addressing climate-related migration. The issue cuts across several areas of international governance – migration and asylum, the environment, development, human rights, and humanitarian aid and assistance – each of which is represented by a number of different United Nations agencies and other institutions. While each institution has a particular mandate (for example, UNHCR for refugees; United Nations Development Programme for development; Office of the High Commissioner for Human Rights (OHCHR) for human rights), none has a sufficiently comprehensive framework to regulate the cross-cutting issues that climate change presents.

UNHCR is the most experienced organisation dealing with displacement, although its mandate primarily concerns refugees and IDPs in conflict situations. It has, however, noted that it is 'keen to influence the necessary dialogue on new or enhanced modalities of international cooperation to develop the capacity of States to respond to the challenges of forced displacement in the context of climate change'.[74] Certainly, UNHCR's long-standing experience would be vital to any institutional developments.

It is nonetheless questionable whether any single organisation could adequately take up this task, in light of the interdisciplinary expertise that is needed to address all aspects of the phenomenon. One option might be an inter-agency approach, with a central coordinating institution. An example of this kind of approach is UNAIDS, whereby a number of institutions feed AIDS-related expertise into a variety of existing United Nations agencies.

Arguably, this is a role already performed informally by the Inter-Agency Standing Committee Task Force on Climate Change, established in 2008 to raise awareness of, advocate for and integrate climate change into various humanitarian agency programs, and to encourage increased inter-agency analysis and cooperation. The Task Force is part of the wider Inter-Agency Standing Committee mechanism, which was created in 1992 in response to a United Nations General Assembly resolution on strengthening humanitarian assistance generally. It comprises representatives from the main United Nations humanitarian agencies, as well as standing invitees. The creation of the Task Force on Climate Change in 2008 is evidence of the adaptability of the international system to respond to new problems, but it remains to be seen how effective it is on displacement issues.

Other interesting alliances have already been forged between organisations whose interests are not traditionally linked. For example, in April 2008 the Climate Change, Environment and Migration Alliance was formed as 'an informal framework for a global multi-stakeholder partnership on climate change, environment and migration'.[75] It is comprised of the International Organization for Migration, Munich Re Foundation, Stockholm Environment Institute, United Nations Environment Programme, OCHA, United Nations University Institute for Environment and Human Security, University of Sussex Development Research Centre on Migration, Globalisation and Poverty, and the World Wildlife Fund.

While this network has no separate international legal identity or mandate from governments, it provides a useful space for sharing ideas and developing common policies. Its main aim of 'mainstream[ing] the environmental and climate change considerations into the migration management policies and practice' perhaps already does informally what a UNAIDS-type structure might do more formally. Whatever form it may take, international coordination will assume greater importance in coming years as climate-related displacement escalates, particularly if carbon emissions are not mitigated sufficiently quickly or comprehensively.

2 International instruments

As mentioned briefly earlier, various calls have been made by some countries, NGOs and academics for a new treaty to deal with climate-related displacement. The nature of these proposals varies widely, from amendments to the 1951 Refugee Convention, to a new instrument specifically defining climate-displaced people and providing for their protection and assistance, to provisions in the UNFCCC, to arrangements for sharing the 'burden' of assisting the displaced.[76]

To date, there has not been a great deal of progress in creating new international standards for responding to climate-related displacement. Neither the United Nations General Assembly nor the Security Council has directly addressed the issue. The United Nations Secretary General's report on the security implications of climate change briefly mentioned displacement.[77] The United Nations Human Rights Council reported on the relationship between climate change and human rights,[78] and adopted a resolution to explore it further.[79] An earlier United Nations report of 2005 considered climate change, disappearing states, human rights and indigenous peoples,[80] and suggested the appointment of an expert to study the problem further (which did not happen). No concrete measures for new international standards have resulted from these initiatives.

The high point of international attention to the problem came at the 2010 Cancún conference, discussed in Chapter 3. The Cancún Adaptation Framework 'invites' states to 'enhance understanding, coordination and cooperation with regard to climate change induced displacement, migration and planned relocation, where appropriate, at national, regional and international levels'.[81] In doing so, states are urged to take into account their 'common but differentiated responsibilities and respective capabilities, and specific national and regional development priorities, objectives and circumstances'.

Although this recommendation does not *require* states to implement migration programs or 'protect' people displaced by climate change, it is an important recognition of the impacts of climate change on human movement and the need for states to address this. Significantly, it recognises migration as a form of adaptation, meaning that adaptation funding could now be directed towards relocation and migration schemes.

Attention to the issue at Cancún is also significant given the failure of countries to agree to deal with it at the Copenhagen climate conference in 2009. There, some groups lobbied to include migration provisions in the Copenhagen Accord,[82] but such provisions were omitted from the final agreement.[83] Australia was opposed to the provisions because it believed that further research was needed on the links between climate change and displacement.[84] Australia may have been wary of any new legal obligations being used by people who may be displaced in the Asia-Pacific region, despite the combined population of Kiribati and Tuvalu being only 110,000 – which is less than Australia's immigration intake in a single year.

Whether the more formal step of pursuing a new international treaty on climate-related displacement is warranted is a difficult question. While there is certainly a need for better international cooperation and responsibility sharing, a new treaty on 'climate refugees' is not necessarily the best way to facilitate this.[85] First, one might query the desirability of a new treaty focusing on a single displacement driver – climate change – rather than poverty or conflict or natural disaster, all of which may cause equal hardship. Displacement may be better addressed in terms of the *needs* of those who move, rather than in terms of what *drives* it.

Secondly, those most at risk of climate-related displacement, such as Pacific Island countries, are divided on whether such an instrument is desirable. Kiribati, for example, is keen for other countries to acknowledge the role of climate change in the precariousness of Kiribati's ongoing existence, by including 'relocation' as an option alongside 'adaptation' and 'mitigation' in international agreements.[86] By contrast, Tuvalu and Micronesia fear that doing so could send the wrong message: industrialised countries might then see displacement as something that can be 'solved' through relocation, without considering the cultural, social and political dimensions of a fair solution. They are also concerned that countries will not reduce carbon emissions if they think their impacts can be addressed in this way. Efforts to secure funding for adaptation and migration options for the future may also be stymied if too much emphasis is placed on the relocation of the whole population. As the Tuvaluan Prime Minister stated: 'We will fight to keep our country, our culture and our way of living. We are not considering any migration scheme. We believe if the right actions are taken to address climate change, Tuvalu will survive'.[87]

Thirdly, and pragmatically, countries presently lack the political will to negotiate a new instrument requiring them to provide protection to additional groups of people. Many countries have 'compassion fatigue' in relation to refugees, economic migrants and other displaced people, fuelled in part by political resistance in domestic politics (such as popular discontent with 'boat people' in Australia, fuelled by some political leaders and media commentators). Even if they did, it would not be easy to compel the ratification, implementation and enforcement of such a treaty. While this does not justify stalling legal developments altogether, it highlights a significant obstacle to a treaty-based solution.

Ultimately, whether or not a new treaty is a good idea depends of course on what is in it. It might be possible, for instance, to craft a relatively open-ended treaty providing for various forms of assistance and cooperation for a range of people displaced by economic, environmental, political, conflict and climate-related reasons. The drawback for countries is that the wider the categories are, the more burdensome and costly are the obligations imposed by the treaty, and the less likely they are to sign up to it. By contrast, a narrow treaty limited to 'climate refugees' risks being both unworkable in practice (since that cause cannot be readily singled out) and unfairly exclusive (since those displaced by 'mere' poverty or conflict, but not climate change, are left out – even if their suffering is more acute).

In light of the difficulties involved in pursuing a new treaty, UNHCR has instead suggested that 'a more viable approach would be to at least develop a global guiding framework for situations of cross-border displacement resulting from climate change and natural disasters'. UNHCR has offered its assistance in developing such a framework, 'which could take the form of temporary or interim protection arrangements'.[88] Some have suggested that the United Nations Guiding Principles on Internal Displacement may provide a useful template. The 2011 Nansen Principles on Climate Change and Displacement also offer overarching principles to guide legal and policy development, developed by key humanitarian agencies and experts.

The advantage of a soft law framework is its flexibility: countries can maintain considerable freedom of action in defining standards of good conduct, while also improving the basis for international cooperation. Based on existing refugee and human rights law principles, such an instrument would not require states to assume new obligations, but rather would clarify how those obligations might apply specifically in the climate change displacement context.

3 Regional frameworks

In addition to the increasing global attention to the problem of climate-related displacement, regional forums and initiatives have also considered the issue. At the Pacific Island Forum in 2008, Pacific leaders (including from Australia and New Zealand) adopted the Niue Declaration on Climate Change, which recognised 'the importance of retaining the Pacific's social and cultural identity, and the desire of Pacific peoples to continue to live in their own countries, where possible'.[89] It also committed the Forum members 'to continue to advocate and support the recognition, in all international fora, of the urgent social, economic and security threats caused by the adverse impacts of climate change and sea level rise to our territorial integrity and continued existence as viable dynamic communities'.[90]

Subsequently, at a high-level Climate Change Conference in Kiribati in November 2010, leaders from 12 countries (Australia, Brazil, China, Cuba, Fiji, Japan, Kiribati, Maldives, Marshall Islands, New Zealand, Solomon Islands and Tonga) adopted the non-binding Ambo Declaration. They expressed their '*[a]larm* at the impacts of the climate change crisis already being felt ... especially the immediate threat to the livelihood and survival of the most vulnerable States', and their support for 'strategies and actions directed at protecting people displaced within or across borders as a result of adverse effects arising from climate change extreme events'.

The non-binding, aspirational Anchorage Declaration of April 2009, adopted at the Indigenous Peoples' Global Summit on Climate Change in Alaska, called on countries to 'recognize, respect and implement the fundamental human rights of Indigenous Peoples', including ensuring that 'Indigenous Peoples have the right to mobility and are not forcibly removed or settled away from their traditional lands and territories'. It also noted that '[i]n the case of climate change migrants, appropriate programs and measures must address their rights, status, conditions, and vulnerabilities'.[91]

Finally, the Council of Europe Parliamentary Assembly's Committee on Migration, Refugees and Population has suggested 'adding an additional protocol to the European Convention on Human Rights, concerning the right to a healthy and safe environment' as a way of 'enhancing the human rights protection mechanisms vis-à-vis the challenges of climate change and environmental degradation processes'.[92] So far, that proposal has not been pursued. While none of these regional initiatives has the binding force of law, they have been significant in raising awareness of, and encouraging dialogue about, the issue, which in turn may prepare the ground for more concrete action over time.

4 The way forward

All of the above developments indicate that, although international and regional institutions are aware of the potential for climate-related displacement, most institutions are still grappling with the problem at a preliminary level and have yet to articulate clear policy responses or make any real commitments in responding to the problem.

A failure to coordinate international responses may well result in ad hoc humanitarian distance becoming the default response of the international community. Such a response is not likely to be adequate in scale, timeliness, durability or comprehensiveness. It risks imposing inequitable obligations on certain countries and would ultimately fail to adequately protect the vulnerable.

There is, accordingly, a need to develop a comprehensive framework for managing new flows of displaced persons, based on burden-sharing principles, a human-rights oriented approach, and the allocation of well-defined institutional responsibilities for humanitarian relief, legal protection and the provision of permanent solutions. This includes a need to develop innovative ways for dealing with the potential disappearance of small island states and extinction of their legal personality, which may carry with it new problems.

For pragmatic reasons, a comprehensive migration strategy need not necessarily involve the formulation of a new international treaty covering, for instance, environmental migration.[93] It would, however, require a commitment by developed states to actively manage the problem, including by providing development assistance for internal relocation where appropriate. In normative terms, it would also demand greater attention to the human rights implications of climate change. Climate change is, however, likely to weaken some states and render them unable to fulfil their obligations to protect and ensure rights.[94] International, regional and national responses to climate-induced displacement (including as a result of conflict) must accordingly be attentive to the fundamental rights of those forced to move.

Increasing labour and economic migration pathways at the national level will also be necessary. Given the political sensitivities of immigration policy in many countries, it may come as no surprise that there is no general international law governing global migration. Migration is generally left to national legal discretion rather than being subject to comprehensive international legal control. But opening up such migration options is one way in which developed countries could proactively plan for and regulate potential displacement, instead of waiting to respond reactively to unpredictable movements in the future. The next section considers some of the limited national legal responses which are already emerging.

F Domestic responses[95]

1 Protection responses

To date, most responses to cross-border climate-related or environmental displacement have been domestic ones rather than international agreements (with the exception of the 2009 Kampala Convention of the African Union, discussed above). They include temporary humanitarian assistance, through schemes such as Temporary Protected Status (TPS) in the United States and potential temporary protection in the European Union, as well as longer lasting refugee-like protection in Sweden, Finland and Switzerland. As mentioned above, it remains to be seen how far complementary protection schemes based on non-return to cruel, inhuman or degrading treatment may be able to encompass flight from habitat destruction or other climatic or environmental problems.

In the United States, the Secretary of Homeland Security may 'designate' a country where there is an ongoing armed conflict threatening people's personal safety, or where:

> (i) there has been an earthquake, flood, drought, epidemic, or other environmental disaster in the state resulting in a substantial, but temporary, disruption of living conditions in the area affected,
> (ii) the foreign state is unable, temporarily, to handle adequately the return to the state of aliens who are nationals of the state, and
> (iii) the foreign state officially has requested designation under this subparagraph. ...[96]

This provides only a temporary status for affected nationals *already* in the United States. For example, in January 2010, as a result of an earthquake in Haiti, the Department of Homeland Security determined an 18-month designation for Haitians who had continuously resided in the United States since 12 January 2010. Haitians unlawfully in the United States, as well as those living there on another visa, were eligible to apply for TPS. A grant of TPS enables the beneficiary to work and precludes deportation for the period of the designation.[97] TPS is not an additional refugee category, however, and it is not available to Haitians who fled the earthquake and arrived in the United States after 12 January 2010.

The European Union Temporary Protection Directive, which has not yet been triggered since its adoption in 2001, was designed to respond to mass influx on account of armed conflict, endemic violence or generalised human rights violations.[98] However, it also has the capacity potentially to cover mass influx in response to climate change impacts.

At the national level in Europe, Swedish asylum law contains a provision extending protection to people who are 'unable to return to the[ir] country of origin because of an environmental disaster'.[99] To date, however, it has never been used. In any case, it is unclear if it would extend to people displaced by climate change, since it seems that it was only ever intended to cover people fleeing specific environmental disasters, such as the Chernobyl nuclear power station accident, rather than broader climate-induced displacement.

Finnish asylum law also provides that a person may be granted asylum if he or she faces a 'threat of death penalty, torture or other inhuman treatment or treatment violating human dignity, or if they cannot return there [to the country of origin] because of an armed conflict or environmental disaster'.[100] Swiss asylum law relating to subsidiary and temporary protection also has the capacity to respond in this way.[101] Again, these laws are untested in relation to climate-related displacement and were not designed as a broad 'catch-all' category.

Prior to being elected to government in late 2007, the Australian Labor Party proposed the creation of a Pacific Rim coalition to accept climate change 'refugees', encouraging the then Australian government to lobby the United Nations to 'ensure appropriate recognition of climate change refugees in existing conventions, or through the establishment of a new convention on climate change refugees'.[102]

Perhaps as a sign of things to come, when Greens Senator Kerry Nettle sponsored the Migration (Climate Refugees) Amendment Bill 2007, which proposed issuing visas to people fleeing 'a disaster that results from both incremental and rapid ecological and climatic change and disruption', the Labor Party was quick to note that its idea of an international response was a *collaborative* approach with other countries, rather than unilateral action by Australia.[103]

Since forming government in late 2007, Labor has not devised any policies on this issue. A Senate inquiry revealed that '[w]hen asked about the possibility of forced re-location from Pacific island countries such as Kiribati and Tuvalu, DFAT [Department of Foreign Affairs and Trade] informed the committee that it was not aware of any government consideration of this matter. Invited to comment again on whether these two islands were under consideration, DFAT replied no'.[104] In November 2009, however, a spokesperson for the then Climate Change Minister, Penny Wong, did acknowledge that permanent migration might ultimately be the only option for some people, which would need to be dealt with by governments in the region.[105]

2 Migration responses

Standard migration schemes may facilitate the movement of people from places hard hit by climate change impacts, even though if not expressly characterised as climate-related programs. The long-term strategy of the government of Kiribati is to secure 'merits-based migration' options to Australia and New Zealand, so that those who want to move have an early opportunity to do so.[106] In this way, the President hopes that 'pockets' of i-Kiribati communities will build up abroad, thereby facilitating the potential relocation of all i-Kiribati over time. The President hopes that this will help to maintain i-Kiribati culture and traditions, and ease his people's adaptation to new cultures and a new way of life.

Australia and New Zealand take different approaches to Pacific migration. Whereas Australia maintains a 'non-discriminatory' policy that does not (formally) privilege any national group, New Zealand is unique in having special concessionary schemes for citizenship and permanent residence for Pacific Islanders, the rationale of which is to promote economic development in Pacific island countries.

New Zealand's Pacific Access Category, established in 2002, is the most recent of these concessionary schemes. Based on an existing scheme for Samoans it replaced previous work schemes and visa waiver schemes for people from Tuvalu, Kiribati and Tonga.[107] It enables an annual quota of 75 citizens each from Tuvalu and Kiribati and 250 from Tonga (and previously Fiji), plus their partners and dependent children, to settle in New Zealand.[108] Eligibility is restricted to applicants between the ages of 18 and 45, who have a job offer in New Zealand, meet a minimum income requirement and have a minimum level of English. Selection is by ballot.

The Pacific Access Category is *not* a response to climate-induced displacement in the Pacific, even though it has been erroneously described as such. Although the scheme was extended to citizens of Tuvalu after a plea from that country's government for special assistance to enable some of its 12,000 citizens to relocate, it is a traditional labour migration program rather than one framed in terms of international protection.

Since 2007, AusAID (the Australian government's foreign aid agency) has funded the Kiribati–Australia Nursing Initiative, which offers around 30 young i-Kiribati the opportunity to train as nurses in Queensland and, after graduation, to remain in Australia. In 2009, Australia implemented a three year Pacific Seasonal Workers Pilot Scheme, modelled in part on New Zealand's Recognised Seasonal Employer scheme.[109] Over three years, up to 2500 visas will be granted to people from Kiribati, Tonga, Vanuatu and Papua New Guinea to work in the Australian horticultural industry for six to seven months in each 12-month period.[110] The program appears to be less successful than anticipated, not least because of a lack of job offers from Australian farmers who seem reluctant to meet minimum labour conditions which can be avoided by relying on 'black market' labour (such as 'backpackers' from Britain, the United States or Europe).

Strategically, Australia and New Zealand would benefit from a more cooperative approach to migration, especially since many Pacific islanders view movement to New Zealand as the first step towards ultimately reaching Australia. Once they obtain New Zealand citizenship,[111] they can freely travel to and work in Australia, as a result of agreements on the free movement of nationals between the two countries.[112]

To summarise, then, the table below sets out the main advantages and disadvantages of the various legal and policy options available for responding to climate-related displacement:

Option	**Advantages**	**Disadvantages**
Treaty	• Global consensus • Universal application • Normative framework • Legally binding	• Lack of political will • Ineffective if not implemented • Problems of definition • May narrow the scope of protection
Guiding Principles	• Based on legal existing obligations • Useful first step which gives states time to develop law over time	• Non-binding (although underlying obligations may be binding) • Effects in practice will vary greatly from country to country
Migration pathways	• Permit planned movement (for slow onset climate change) • Discretionary and flexible • Recognise human agency and choice • Migration eases population pressure • Create diaspora communities (facilitating future movement) • Remittances flow to country of origin • Fill jobs in receiving states	• Discretionary character limits ability of all persons at risk to move abroad • Absence of binding protection against *non-refoulement* to harm • Risk of labour exploitation • Migration may only be temporary • Migration may preclude family unity

Option	Advantages	Disadvantages
Group relocation	• An option for vulnerable groups • Keeps communities together • May help to preserve group identity	• Balancing needs of existing population with relocated communities • Break with land, culture, history • Loss of identity • Potential destruction of self-determination and national sovereignty

G Conclusion

There remains a clear need to better define legal and institutional frameworks for responding to climate-related displacement. However, as this chapter has shown, the task is a complex one because of the multiple types of movement that will occur. Following the Copenhagen conference in 2009 and Cancún in 2010, it seems that any international consensus on climate-related issues will be difficult to achieve.

This is not necessarily detrimental to the migration dimension of climate change, given that existing research on affected populations consistently highlights the importance of local factors in determining appropriate responses. While reaching agreement on overarching issues, such as the need to cooperate and respond to adaptation needs, is fundamental, the detail and implementation of responses will need to be crafted for particular circumstances.

Arguably, it is more appropriate and culturally sensitive to respond on a regional or even bilateral basis. This enables the special needs and characteristics of the threatened population to be taken into account, in determining who should move, when, in what fashion and with what outcomes. Enhanced opportunities for structured economic and labour migration should be part of the solution, alongside natural disaster relief responses to specific events. Legal frameworks such as international human rights law play a fundamental role in determining how such responses should be developed, but achievement of an international treaty specifically on climate change migrants is unlikely to yield many additional practical or legal benefits.

However difficult it is to determine the best way forward, countries will ultimately need to develop responses that acknowledge the need for cross-border movement in certain circumstances and which regularise the status of those who move, either through humanitarian or migration schemes. A 'wait and see' approach will not be enough.

Notes

1 That is not to say that 'natural' disasters are without human causes: see, eg, Wisner, B, Blaikie, P, Cannon, T and Davis, I, 2004, *At Risk: Natural Hazards, People's Vulnerability and Disasters*, 2nd edn, Routledge, London, which argues that few disasters are ever 'natural'; they are a combination of environmental *plus* socio-economic and political factors.

2 President of the Global Humanitarian Forum, Kofi Annan, described 'millions of people' being 'uprooted or permanently on the move as a result' of climate change, with '[m]any more millions' to follow: Global Humanitarian Forum, 2009, *The Anatomy of a Silent Crisis*, Geneva, p ii, available at <http://ghfgeneva.org/Portals/0/pdfs/human_impact_report.pdf>. Debates about numbers remain highly contentious: see, eg, Kniveton, D et al, 2008, *Climate Change and Migration: Improving Methodologies to Estimate Flows*, International Organization for Migration (IOM), Migration Research Series No 33, Geneva; Castles, S, 2002, 'Environmental Change and Forced Migration: Making Sense of the Debate', *New Issues in Refugee Research* Working Paper No 70, UNHCR, Geneva.

3 Hegerl, GC et al, 'Understanding and Attributing Climate Change' in Solomon, S et al (eds), 2007, *Climate Change 2007: The Physical Science Basis: Contribution of Working Group I to the Fourth Assessment Report of the Intergovernmental Panel on Climate Change*, Cambridge University Press, Cambridge; Intergovernmental Panel on Climate Change (IPCC), 2007, *Climate Change 2007: Synthesis Report: Summary for Policymakers*, Cambridge University Press, Cambridge; Stern, N, 2007, *The Economics of Climate Change: The Stern Review*, Cambridge University Press, Cambridge, pp vi–ix.

4 Mimura, N et al, 'Small Islands' in Parry, ML et al (eds), 2007, *Climate Change 2007: Impacts, Adaptation and Vulnerability. Contribution of Working Group II to the Fourth Assessment Report of the Intergovernmental Panel on Climate Change*, Cambridge University Press, Cambridge, p 689.

5 See, eg, Myers, N, 1993, 'Environmental Refugees in a Globally Warmed World' 43 *BioScience* 752; Myers, N, 2005, 'Environmental Refugees: An Emergent Security Issue', 13th Economic Forum, Prague; cf Black, R, 2001, 'Environmental Refugees: Myth or Reality?', *New Issues in Refugee Research* Working Paper No 34, UNHCR, Geneva; Castles, above n 2. See also Brown, O, 2008, 'The Numbers Game' 31 *Forced Migration Review* 8.

6 Kniveton et al, above n 2, p 32.

7 Myers, 1993, above n 5; Myers, N, 'Environmentally-Induced Displacements: The State of the Art' in IOM/UNHCR, 1996, *Environmentally-Induced Population Displacements and Environmental Impacts Resulting from Mass Migration*, IOM/UNHCR International Symposium, Geneva; Myers, N, 1997, 'Environmental Refugees' 19 *Population and Environment* 167; Myers, N, 2002, 'Environmental Refugees: A Growing Phenomenon of the 21st Century' 357 *Philosophical Transactions of the Royal Society: Biological Sciences* 609; Myers, 2005, above n 5.

8 Christian Aid, 2007, *Human Tide: The Real Migration Crisis*, London.

9 Ibid, pp 5–6.

10 Stern, above n 3.

11 Annan, above n 2, p ii.

12 United Nations High Commissioner for Refugees, António Guterres, in an interview with *The Guardian* (reported by Julian Borger, 17 June 2008), available at <http://www.guardian.co.uk/environment/2008/jun/17/climatechange.food>.

13 Kälin, W, 'Conceptualising Climate Change' in McAdam, J (ed), 2010, *Climate Change and Displacement: Multidisciplinary Perspectives*, Hart Publishing, Oxford, p 81. See generally Parry et al, above n 4, chs 9, 10 and 16.

14 EACH-FOR, 2009, *Synthesis Report*, 72, available at <http://www.each-for.eu/documents/EACH-FOR_Synthesis_Report_090515.pdf>; see also Laczko, F and Aghazarm, C (eds), 2009, *Migration, Environment and Climate Change: Assessing the Evidence*, IOM, Geneva.

15 EACH-FOR, above n 14, p 5.

16 See, eg, 'Carteret Islanders Become First Climate Refugees: PNG Relocates Families as Island Home Disappears', *Pacific Island Report*, 4 May 2009, available at <http://archives.pireport.org/archive/2009/may/05-04-09.htm>, reporting a 1 May 2009 story from the *Papua New Guinea Post-Courier*.

17 Fleming, M, 16 December 2009, 'Climate Change Could Become the Biggest Driver of Displacement: UNHCR Chief', citing the UN High Commissioner for Refugees, António Guterres, available at <http://www.unhcr.org/4b2910239.html>.

18 Inter-Agency Standing Committee Working Group on Migration/Displacement and Climate Change, 20 September 2008, *Displacement and Climate Change: Towards Defining Categories of Affected Persons*, First Draft of a Working Paper, Geneva.

19 Warner, K, Dun, O and Stal, M, 2008, 'Field Observations and Empirical Research' 31 *Forced Migration Review* 13 at 14.

20 Ibid, p 13.

21 Kälin, above n 13.

22 Bronen, R, 2008, 'Alaskan Communities' Rights and Resilience' 31 *Forced Migration Review* 30 at 32.

23 Ibid, pp 30–31.

24 Paskal, C, 2007, *How Climate Change is Pushing the Boundaries of Security and Foreign Policy*, Chatham House Briefing Paper, London, p 4, citing modelling published by the UNEP, University of Dhaka, World Bank et al.

25 Mimura et al, above n 4, pp 692–93.

26 Author's interview with Kiribati Solicitor-General David Lambourne, 8 May 2009, Kiribati.

27 Kälin, W, 16 July 2008, 'The Climate Change – Displacement Nexus', ECOSOC Panel on Disaster Risk Reduction and Preparedness: Addressing the Humanitarian Consequences of Natural Disasters', para 4, available at <http://www.brookings.edu/speeches/2008/0716_climate_change_kalin.aspx>.

28 See remarks made by the UN High Commission for Refugees António Guterres in an interview with *The Guardian*, above n 12.

29 Van der Geest, K and de Jeu, F, 2008, 'Ghana' 31 *Forced Migration Review* 16 at 16.

30 Ibid.

31 Ibid.

32 Ibid.

33 McAdam, J and Saul, B, 2010, 'Displacement with Dignity: Climate Change, Migration and Security in Bangladesh' (2010) 53 *German Yearbook of International Law* 233.

34 See, eg, Barnett, J and Webber, M, 'Migration as Adaptation: Opportunities and Limits' in McAdam (ed), above n 13.

35 Morrissey, J, 2008, 'Rural-Urban Migration in Ethiopia' 31 *Forced Migration Review* 28 at 29.

36 Ibid, p 29 (emphasis added).

37 See Betts, A and Kaytaz, E, 2009, 'National and International Responses to the Zimbabwean Exodus: Implications for the Refugee Protection Regime', *New Issues in Refugee Research*, Research Paper No 175, UNHCR, Geneva.

38 Convention relating to the Status of Refugees (adopted 28 July 1951, entered into force 22 April 1954) 189 UNTS 137 ('Refugee Convention'), Article 1A(2), read in conjunction with Protocol relating to the Status of Refugees (adopted 31 January 1967, entered into force 4 October 1967) 606 UNTS 267.

39 See Goodwin-Gill, GS and McAdam, J, 2007, *The Refugee in International Law*, 3rd edn, Oxford University Press, Oxford, p 92.

40 Ibid, pp 79–80; *Applicant A v Minister for Immigration and Ethnic Affairs* (1997) 190 CLR 225 at 341 (Dawson J).

41 NZ cases: *Refugee Appeal No 72719/2001*, RSAA (17 September 2001) (Tuvalu); *Refugee Appeal No 72313/2000*, RSAA (19 October 2000) (Tuvalu); *Refugee Appeal No 72314/2000*, RSAA (19 October 2000) (Tuvalu); *Refugee Appeal No 72315/2000*, RSAA (19 October 2000) (Tuvalu); *Refugee Appeal No 72316/2000*, RSAA (19 October 2000) (Tuvalu); *Refugee Appeal Nos 72179–72181/2000*, RSAA (31 August 2000) (Tuvalu); *Refugee Appeal Nos 72189–72195/2000*, RSAA (17 August 2000) (Tuvalu); *Refugee Appeal No 72185/2000*, RSAA (10 August 2000) (Tuvalu); *Refugee Appeal No 72186/2000*, RSAA (10 August 2000) (Tuvalu). Australian cases: *1004726* [2010] RRTA 845 (30 September 2010) (Tonga); *0907346* [2009] RRTA 1168 (10 December 2009) (Kiribati); *N00/34089* [2000] RRTA 1052 (17 November 2000) (Tuvalu); *N95/09386* [1996] RRTA 3191 (7 November 1996) (Tuvalu); *N96/10806* [1996] RRTA 3195 (7 November 1996) (Tuvalu); *N99/30231* [2000] RRTA 17 (10 January 2000) (Tuvalu); *V94/02840* [1995] RRTA 2383 (23 October 1995) (Tuvalu).

42 *SERAC and CESR v Nigeria, Decision regarding Communication No 155/96*, ACHPR/COMM/A044/1; *15th Annual Activity Report of African Commission on Human and Peoples' Rights*, 7 May 2002, annex V, para 44.

43 For a detailed analysis, see McAdam, J, 2010, '"Disappearing States", Statelessness and the Boundaries of International Law' in McAdam (ed), above n 13.

44 Weis, P, 1979, *Nationality and Statelessness in International Law*, 2nd rev edn, Sijthoff and Noordhoff International Publishers, Alphen aan den Rijn, p 136.

45 Convention relating to the Status of Stateless Persons (adopted 28 September 1954, entered into force 6 June 1960) 360 UNTS 117 ('1954 Convention'), Article 1(1).

46 See Final Act appended to Convention on the Reduction of Statelessness (adopted 30 August 1961, entered into force 13 December 1975) 989 UNTS 175 ('Convention on the Reduction of Statelessness').

47 Refugee Convention, Article 32.

48 See, eg, UNGA Res 50/152 (9 February 1996), reiterated in UNGA Res 61/137 (25 January 2007), UNGA Res 62/124 (24 January 2008), UNGA Res 63/148 (27 January 2009).

49 Though poorly ratified and implemented, the 1961 Convention on the Reduction of Statelessness obliges states to ensure that any transfer of territory does not render people stateless: Convention on the Reduction of Statelessness, Article 10.

50 Campbell, J, 'Climate-Induced Community Relocation in the Pacific: The Meaning and Importance of Land' in McAdam (ed), above n 13.

51 See *Case concerning Phosphate Lands in Nauru (Nauru v Australia)* [1992] ICJ Rep 240.

52 'Haitians to Africa? Senegal Resettlement Plans Gain Steam', 2 February 2010, *Christian Science Monitor*, available at <http://www.csmonitor.com/layout/set/print/content/view/print/277552>.

53 'African Union to Consider "Land for Haitians" Plan', 31 January 2010, *Reuters*, available at <http://www.reuters.com/assets/print?aid=USTRE60U0IV20100131>.

54 Ibid.

55 'Haitians to Africa?', above n 52.

56 Asia Pacific Forum, *Human Rights and the Environment: Background Paper*, APF 12, Sydney, p 26.

57 *Case Concerning the Gabčikovo-Nagymaros Project (Hungary v Slovakia)* 1997 ICJ 92 (Separate Opinion of Judge Weeramantry), para A(b).

58 *Report of the Office of the United Nations High Commissioner for Human Rights on the Relationship between Climate Change and Human Rights*, UN Doc A/HRC/10/61, 15 January 2009.

59 Human Rights and Equal Opportunity Commission (HREOC) (Australia), 2008, *Background Paper: Human Rights and Climate Change*, HREOC, Sydney.

60 Petition to the Inter-American Commission on Human Rights Seeking Relief from Violations resulting from Global Warming caused by Acts and Omissions of the United States, 7 December 2005, available at <http://www.earthjustice.org/library/legal_docs/petition-to-the-inter-american-commission-on-human-rights-on-behalf-of-the-inuit-circumpolar-conference.pdf>.

61 *Maya Indigenous Communities of the Toledo District (Belize Maya)* Case 12.053 Inter-American Commission on Human Rights (2004), para 120.

62 See Inter-American Commission on Human Rights, *Report on the Situation of Human Rights of a Segment of the Nicaraguan Population of Miskito Origin* 76, 1983, OEA/Ser.L/V/II.62, Doc 10, rev. 3, para II.B.15.

63 *D v United Kingdom* (1997) 24 EHRR 423; *N v United Kingdom* [2008] ECHR 453; *HLR v France* (1997) 20 EHRR 29 at para 42; see also the views of Committee against Torture, as in *AD v The Netherlands*, Communication No 96/1997 (24 January 2000), UN Doc CAT/C/23/D/96/1997, para 7.2. See discussion in Goodwin-Gill and McAdam, above n 39, pp 350–51.

64 *R v Special Adjudicator ex parte Ullah* [2004] UKHL 26; Human Rights Committee, 11 April 1986, *General Comment 15: The Position of Aliens under the Covenant*, para 5; see also Human Rights Committee, 10 November 1989, *General Comment 18: Non-Discrimination*.

65 African Union Convention for the Protection and Assistance of Internally Displaced Persons in Africa (adopted 22 October 2009, not yet in force) ('Kampala Convention').

66 Kampala Convention, Article 5(4).

67 Cashman, P and Abbs, R, 'Liability in Tort for Damage arising from Human-Induced Climate Change' in Lyster, R (ed), 2010, *In the Wilds of Climate Law*, Australian Academic Press, Brisbane, p 245.

68 Ibid, p 245, citing Durrant, N, 2007, 'Tortious Liability for Greenhouse Gas Emissions? Climate Change, Causation and Public Policy Consideration' 7 *Queensland University of Technology Law and Justice Journal* 404 at 406. See *Agar v Hyde* (2000) 201 CLR 552 at paras 66-67 (Gaudron, Gummow and Hayne JJ).

69 Cashman and Abbs, above n 67, p 245.

70 Ibid, p 248, referring to Durrant, above n 68, p 411.

71 Ibid, p 249 referring to Smith, J and Shearman, D, 2006, *Climate Change Litigation: Analysis of the Law, Scientific Evidence and Impacts on the Environment, Health and Property*, Presidian Legal Publications, Adelaide, p 107.

72 Cashman and Abbs, above n 67, p 250, referring to Mank, B, 'Civil Remedies' in Gerrard, MB (ed), 2008, *Global Climate Change and U.S. Law*, American Bar Association, Chicago, p 201.

73 On the question of causation and state responsibility, see Verheyen, R, 2006, *Climate Change Damage and International Law: Prevention Duties and State Responsibility*, Martinus Nijhoff Publishers, Leiden/Boston, ch 5.

74 *UNHCR Factsheet*, October 2009. See also UNHCR, 'Climate Change, Natural Disasters and Human Displacement: A UNHCR Perspective' (23 October 2008, revised 14 August 2009).

75 See *About the Climate Change, Environment and Migration Alliance*, available at <http://www.ccema-portal.org/>.

76 For a discussion of all the various proposals, see McAdam, J, 2011, 'Swimming against the Tide: Why a Climate Change Displacement Treaty is Not *the* Answer' (2011) 23 *International Journal of Refugee Law* 1.

77 *Climate Change and Its Possible Security Implications: Report of the Secretary-General*, UN Doc A/64/350, 11 September 2009, paras 54–63 and Box III.

78 See, eg, OHCHR, 15 January 2009, *Report of the Office of the High Commissioner for Human Rights on the Relationship Between Climate Change and Human Rights*, UN Doc A/HRC/10/61.

79 Human Rights Council, 25 March 2009, *Human Rights and Climate Change*, Res 10/4, 41st mtg.

80 10 August 2005, Res 2005/20. No such reports appear to have been compiled: UN, 12 August 2008, *Human Rights Council Advisory Committee Discusses Studies Prepared by the Sub-Commission on the Promotion and Protection of Human Rights*, Office of News and Media, Geneva, available at <http://www.unog.ch/80256EDD006B9C2E/(httpNewsByYear_en)/71817209AFDCFC25C12574A3003A2068?OpenDocument>.

81 Cancún Conference, *Outcome of the Work of the Ad Hoc Working Group on Long-Term Cooperative Action under the Convention*, Draft Decision -/CP.16, December 2010, para 14, available at <http://unfccc.int/files/meetings/cop_16/application/pdf/cop16_lca.pdf#page=3>.

82 See Negotiating Text, UN Doc FCCC/AWGLCA/2009/14, 20 November 2009, paras 12(c) and 13(b), available at <http://maindb.unfccc.int/library/view_pdf.pl?url=http://unfccc.int/resource/docs/2009/awglca7/eng/14.pdf>; Sekoli, B, Chair of the Least Developed Countries (LDCs) group at Copenhagen cited in IRIN (Humanitarian News and Analysis), 11 December 2009, 'Good News for Climate Change Migrants', available at <http://www.irinnews.org/Report.aspx?ReportId=87405>.

83 Copenhagen Accord (adopted 18 December 2009), available at <http://unfccc.int/files/meetings/cop_15/application/pdf/cop15_cph_auv.pdf>.

84 Lawton, C, 17 December 2009, 'What about Climate Refugees? Efforts to Help the Displaced Bog Down in Copenhagen', *Spiegel Online*, available at <http://www.spiegel.de/international/europe/0,1518,druck-667256,00.html>.

85 For a detailed analysis of this, see McAdam, above n 76.

86 Author's interview with President Anote Tong, 12 May 2009, Kiribati.

87 European Parliament, 10 December 2009, 'Prime Minister Apisai Ielemia Says Climate Change Threatens Tuvalu's Survival', Press Release, Strasbourg, available at <http://www.egovmonitor.com/node/31980/print>, quoting the Tuvaluan Prime Minister. See similar comments made by the government of Nauru when it was proposed that its population relocate to Australia in the 1960s, in McAdam, above n 43.

88 'Statement by António Guterres, United Nations High Commissioner for Refugees' (Nansen Conference on Climate Change and Displacement in the 21st Century, Oslo, 6 June 2011), p 4; see similarly 'Chairperson's Summary' (Nansen Conference on Climate Change and Displacement in the 21st Century, Oslo, 6–7 June 2011), para 23.

89 Pacific Island Forum, 2008, Niue Declaration on Climate Change, Preamble.

90 Ibid, operative part.

91 Indigenous Peoples' Global Summit on Climate Change, 2009, Anchorage Declaration, para 11.

92 Council of Europe (COE) Parliamentary Assembly, Committee on Migration, Refugees and Population, 23 December 2008, *Environmentally Induced Migration and Displacement: A 21st Century Challenge*, COE Doc 11785, paras 6.3 and 121 respectively.

93 As suggested by the German Advisory Council on Global Change: Summary for Policy Makers, 'World In Transition: Climate Change as a Security Risk', May 2007, 11.

94 McAdam and Saul, above n 33.

95 For a fuller discussion, see McAdam, J, 2011, *Climate Change Displacement and International Law: Complementary Protection Standards*, UNHCR Legal and Protection Policy Research Series, PPLA/2011/03.

96 INA § 244, 8 USC § 1254.

97 US Citizenship and Immigration Services, last updated 1 February 2010, *Temporary Protected Status – Haiti*, available at <http://www.uscis.gov/portal/site/uscis/menuitem.eb1d4c2a3e5b9ac89243c6a7543f6d1a/?vgnextchannel=e54e60f64f336210VgnVCM100000082ca60aRCRD&vgnextoid=e54e60f64f336210VgnVCM100000082ca60aRCRD>.

98 Council Directive 2001/55/EC of 20 July 2001 on Minimum Standards for Giving Temporary Protection in the Event of a Mass Influx of Displaced Persons and on Measures Promoting a Balance of Efforts between Member States in Receiving Such Persons and Bearing the Consequences Thereof [2001] OJ L212/12, Article 2(c).

99 Swedish Aliens Act, ch 4, s 2(3).

100 Finnish Aliens Act 301/2004, ss 88 (residence permits) and 109 (temporary protection in cases of mass displacement).

101 Kälin, above n 13, p 100, noting the conclusion of an inter-departmental round table discussion by the Swiss Ministry of Foreign Affairs on 13 January 2009.

102 Australian Labor Party (ALP), 2006, *Our Drowning Neighbours: Labor's Policy Discussion Paper on Climate Change in the Pacific*, ALP, Canberra, p 10.

103 Commonwealth of Australia, 9 August 2007, *Senate: Official Hansard*, Canberra, p 100ff.

104 Senate Foreign Affairs, Defence and Trade References Committee, November 2009, *Economic Challenges facing Papua New Guinea and the Island States of the Southwest Pacific*, Commonwealth of Australia, Canberra, para 6.60.

105 Morton, A, 21 November 2009, 'Land of the Rising Sea', *Sydney Morning Herald*, Sydney, available at <http://www.smh.com.au/environment/land-of-the-rising-sea-20091120-iqub.html>.

106 Interview with President; see also Goering, L, 9 December 2009, 'Kiribati Officials Plan for "Practical and Rational" Exodus from Atolls', *Reuters AlertNet*, available at <http://www.alertnet.org/db/an_art/60714/2009/11/9-181804-1.htm>, referring to remarks by i-Kiribati officials at Copenhagen.

107 Dalziel, L, 20 December 2001, 'Government Announces Pacific Access Scheme', New Zealand Government, Wellington, available at <http://www.beehive.govt.nz/node/12740>.

108 Immigration New Zealand, last updated 19 March 2010, *Pacific Access Category*, available at <http://www.immigration.govt.nz/migrant/stream/live/pacificaccess/>.

109 See, eg, Ramasamy, S, Krishnan, V, Bedford, R and Bedford, C, 2008, 'The Recognised Seasonal Employer Policy: Seeking the Elusive Triple Wins for Development through International Migration' 23 *Pacific Economic Bulletin* 171, available at <http://peb.anu.edu.au/pdf/PEB_23_3_Ramasamy%20et%20al_WEB.pdf>; New Zealand Department of Labour, *Labour Market Strategy: Recognised Seasonal Employer Policy*, available at <http://www.dol.govt.nz/initiatives/strategy/rse/index.asp>.

110 See Australian Government, Department of Education, Employment and Workplace Relations, 2010, *Pacific Seasonal Worker Pilot Scheme Home*, available at <http://www.deewr.gov.au/Employment/Programs/PSWPS/Pages/default.aspx>.

111 For people who obtained permanent residence post-21 April 2005, the waiting period is five years. Before that time, it is three years: New Zealand Government, 2010, *Citizenship*, available at <http://www.teara.govt.nz/en/citizenship/3>.

112 1973 Trans-Tasman Travel Arrangement; see also the Special Category Visa (SCV) for New Zealand citizens since 1994.

Chapter 6

Climate Wars? Conflict, Security and Climate Change

Key Issues

- The political, economic and environmental instability associated with climate change may stimulate violence or conflict, or exacerbate existing causes of such conflict. Climate change is more likely to provoke social conflicts and affect 'human security' (that is, basic rights to be free from poverty or hunger) than to trigger traditional military violence between governments, although conventional threats to national security cannot be ruled out.
- Climate change is likely to aggravate the scarcity of basic resources such as food, water and productive land, by changing rainfall or water availability, increasing the frequency and severity of extreme weather events, accelerating desertification and causing the spread of pests and diseases. Such processes intensify competition over access to increasingly scarce resources within and between countries, but especially in weaker developing countries.
- Some attempts to combat climate change – especially diverting food crops into bio-fuels – have had the unintended effect of decreasing food production and worsening food shortages. In the long term, more rational and equitable mechanisms for sharing scarce food resources across countries are needed, starting by dismantling agricultural trade protectionism.
- Climate change will aggravate water scarcity, potentially intensifying existing transboundary conflicts over rivers in South Asia, Southeast Asia and the Middle East, and internal conflict in Latin America and even in stable countries such as Australia. There is an urgent need to develop cooperative water management strategies within and between countries.

- Food and water shortages interact with other effects of climate change to adversely affect public health and weaken political and economic governance in some developing countries.
- Climate-induced displacement may also encourage tensions within and between countries. In Bangladesh, for instance, outright warfare is an unlikely consequence compared to the risk of lower level social conflicts arising from how forced migrants are received by host communities and local governments.
- Climate change may sharpen disputes between states over maritime boundaries and economic rights in the seas and polar regions. Despite sabre-rattling by some Arctic nations, military conflict seems unlikely to occur.
- Given that an international agreement will come too late to neutralise all climate-related security threats, it is necessary to consider how other areas of international law and global institutions – including those offering humanitarian aid – might prevent and respond to those security risks.

A Introduction

Alarming predictions have been made about the potential for climate change to fuel 'climate wars' and other violent conflicts. At one extreme, it has been suggested that the uncontrolled effects of climate change could generate violent conflicts on the scale of the world wars – but lasting for centuries.[1] Another study suggests that 46 countries with 2.7 billion people are at high risk of conflict due to climate change interacting with other causes of conflict, while a further 56 countries with 1.2 billion people are at high risk of political instability possibly leading to conflict in the long term.[2]

Growing concern about the security risks posed by climate change has emerged in major global institutions such as the United Nations and the European Union.[3] In 2007, a special meeting of the United Nations Security Council discussed the security implications of climate change, with the United Kingdom suggesting that the mass violence in Darfur, Sudan, was partly caused by climate change.[4] In 2008, the UN General Assembly expressed concern that 'the adverse impacts of climate change, including sea-level rise, could have possible security implications', while in 2009 the UN Secretary-General issued a detailed report on the issue.[5]

National militaries too (including the Australian Defence Force) have begun to view climate change as a national security threat,[6] whether because of rising international competition over scarce resources, the threat of domestic instability due to food riots or extremists potentially reacting against the adverse social effects of climate change.

It is too simple, however, to claim that climate change will trigger global 'climate wars'.[7] Not only is there little evidence for any direct causal connection,[8] it is clear that human conflicts typically result from a complex combination of causes. The conflict in Darfur is a good example. Climate change has probably reduced rainfall in Darfur by 30 per cent over the past 40 years, contributing to prolonged droughts in the 1960s and 1970s.[9] The droughts in turn affected the productivity of land and increased tensions between different ethnic groups over scarce resources.

Yet, neighbouring African countries such as Mali and Chad experienced similar climatic changes (including desertification) without erupting in comparable conflict, mainly because other underlying causes of conflict were not present there as in Sudan. In the first place, the political situation in Sudan over time had involved increasing centralisation of power, serious inequalities in economic wealth and political participation, unfair land redistribution policies (some of it encouraged by misguided programs of the World Bank) and a lack of respect for the human rights of minorities.[10]

These factors disenfranchised tribal groups in Darfur and aggravated historical grievances, exclusionary, ethnicity-based politics and resentment towards the government. As tensions increased because of the additional resource scarcities brought by climate change, other factors came into play which tended to war: the proliferation of cheap weapons, the militarisation of young people, the recruitment of militias (*janjaweed*) motivated in large part by self-interest and the destruction of traditional social structures for mitigating conflict.

The violence in Darfur was thus not inevitably impelled by climate change, but resulted from an array of contributory factors.

The Sudan example suggests that humans ultimately choose to wage war, and that some political and social structures are better able to defuse or deal with it, even if climate change can make peace more difficult.

Climate change clearly has the potential to multiply[11] other underlying causes of conflict, whether competition over scarce resources, political instability, ethnic politics, sectarian violence and so on. One study by a think tank, International Alert, identified the following climate change related factors as potentially contributing to climate change:[12]

Physical consequences of climate change	**Knock-on socio-political consequences of climate change**
• Melting glaciers • Sea-level rise • Loss of island coastline • Less usable land • Droughts • Floods • Desertification • Spread of disease and pestilence • Changes to crop seasons and output	• Livelihood insecurity • Food insecurity • Increased social tension • Less access to useable water • Decreased trade • Decline in human health • Increased poverty • Decreased physical security • Increased migration

This chapter suggests that climate change may affect the security of human communities in a variety of ways. It is more likely to affect 'human security'[13] (that is, basic rights to be free from poverty and violence) and create social conflicts than to provoke traditional military violence between governments, although conventional threats to national security cannot be ruled out in exceptional cases.

More specifically, climate change is likely to aggravate the scarcity of basic resources such as food, water and productive land, which are essential to human dignity, thus increasing intra- and inter-state competition over access to increasingly scarce resources. Such shortages may interact with other effects of climate change to impact adversely on public health and to weaken political and economic governance in some countries, leading to social instability and possible social conflicts. Climate-induced displacement of the kind discussed in the previous chapter may also encourage tensions within and between countries. Finally, climate change may sharpen disputes between states over maritime boundaries and economic rights in the seas and polar regions, some of which may involve a risk of escalation into military confrontation.

Any post-Kyoto global compact for reducing carbon emissions and mitigating climate change will come too late to neutralise all the security risks flowing from the climate change which has already occurred, or is underway and irreversible in coming decades.[14] While such an agreement would be an important part of containing those risks, it is also necessary to consider how other areas of international law and global institutions might prevent and respond to the risks.

B Food insecurity and conflict

In 2007–08, 'food riots' in response to escalating food prices affected many developing countries and were widely seen to be connected with climate change. People were killed in violent demonstrations in Cameroon, Egypt, Haïti, Mozambique, Somalia and Yemen, while violence occurred in Bangladesh, Burkina Faso, Côte d'Ivoire, India, Mauritania and Senegal. Protests or demonstrations also took place in Bolivia, Indonesia, Morocco, Pakistan, Sri Lanka, South Africa, Thailand and Uzbekistan. Police and even the military were called upon in various countries to restore order.

Though far from 'climate wars', the food riots signal one kind of destabilising social conflict that can erupt if the effects of climate change are not well managed on a global scale. While food insecurity has long been linked to social conflict, climate change potentially aggravates the risks by affecting global food supply in a number of ways. (Chapter 2 already discussed the impact of climate change on Australian agriculture.)

On land, agricultural production is sensitive to rising temperatures, changes in rainfall or water availability (including due to shifts in seasonal monsoon patterns and reductions in glacial melt-water) and extreme weather events (such as storms, floods, droughts). In the past two decades alone, there was an increase from 200 to 400 major disasters per year.[15] Increasing aridity and desertification are climate-related factors affecting agriculture, along with movements in insects and diseases affecting crops and livestock.

In coastal areas, sea-level rises have inundated agricultural land and increased salinity. In Asia, a sea-level rise of less than one metre could destroy half of Bangladesh's rice-growing land, while much of the Mekong Delta in Vietnam would be inundated.[16] At sea, changes in sea-levels, temperatures and currents, and ocean acidification,[17] are aggravating already stressed marine resources and fish stocks (an important protein source for many coastal populations). In the Pacific, for instance, overfishing of some species by Asian countries such as South Korea, China, Taiwan and Japan is currently jeopardising the food security and economic interests of small Pacific Island countries.

The price of staple foods increased by 80 per cent over the three years from 2006 to 2008.[18] It is estimated that over 100 million people were pushed into poverty in 2007–08 alone due to rising food (and energy) prices,[19] taking the total number of malnourished people worldwide from 850 to 963 million. This also rapidly undermined the progress that had been made in addressing poverty under the Millennium Development Goals (which are non-binding commitments by governments to meet human development targets). The World Bank estimates that 33 countries are at risk of political destabilisation and internal conflict due to food price inflation,[20] while 36 countries with 1.4 billion people are at risk of food or water scarcity by 2025.[21]

In India, for example, rice and wheat production has declined in recent years due to temperature increases, affecting food security in an agriculturally dependent and underdeveloped country. The failure of the monsoon in 2001 led to a 19 per cent decrease in rainfall and reduced India's Gross Domestic Product (GDP) by 3 per cent in that year.[22] Even developed countries such as the United Kingdom, which relies heavily on food imports, may face food shortages as climate change affects global supply, in circumstances where domestic agricultural production is affected by higher temperatures, changing rainfall, and new crop and animal diseases.[23] The drought which affected the Australian wheat crop in 2008 contributed to a 130 per cent rise in the global wheat price, which flowed on to a 15 per cent increase in food prices overall in Britain.

While some countries such as Russia and Canada may benefit from longer growing seasons and become more productive, globally such gains would be offset by declines in much of the developing world, such as Africa and Asia, and in food-producing developed countries such as Australia and the United States.

While there will undoubtedly be advances in food science and bio-technology, a new 'green revolution' – such as that of the 1960s – appears unlikely given present trends. Without greater research into adapting crops for climate change, it is estimated that wheat yields may decline by 30 per cent in developing countries, resulting in price rises of almost 200 per cent by 2050.[24] The Food and Agriculture Organization estimates that crop yields could decrease by 25–35 per cent if the global temperature rises by 3–4°C.[25]

If the effects of climate change on agriculture intensify over time, food scarcity is likely to increasingly generate social conflicts or even more intense violence over scarce food resources. Particularly at risk are those developing countries which are heavily dependent on food imports. Algeria, for example, imports 45 per cent of its food while only 3 per cent of its land area is arable, rendering it vulnerable to global food price shocks, climate change related water scarcity, and desertification in the Saharan region[26] – and to the political radicalisation that has frequently exposed Algerians to violence in the past.

The world's population is estimated to grow from 6.8 billion to 9 billion people over the next 40 years, placing even greater pressure on agricultural production. As one scientist observes: 'We are going to have to produce as much food in the next 50 years as was produced over the past 5,000 years'.[27] Food scarcity may be aggravated by increasingly high living standards in developing countries, which brings changes in consumer diets among the new middle classes – particularly more demand for meat, thus diverting grain supplies from staple foods to livestock feed – placing further pressure on food prices.

1 Other factors contributing to food insecurity

Climate change alone is not wholly responsible for the 'food riots' and protests of recent years. In some places, the riots were also partly a response to local factors such as repressive governments, decision-making about agricultural policy or even exploitative labour standards. While understandable, some national responses to the emerging food crisis in 2007–08 tended to aggravate the rise in global food prices, as when some governments controlled basic food prices in their own countries or restricted the export of staple foods. In 2008, for instance, Kazakhstan's limits on wheat exports, designed to protect its own food security, increased the global wheat price by 25 per cent in one day.[28]

A range of other global conditions contributed to higher food prices and the social conflicts which followed. Higher energy prices (and related increases in fertiliser and transport costs, and currency fluctuations) may explain up to 25–30 per cent of the international food price increases from January 2002 to June 2008.[29] The global financial crisis of 2008–09 squeezed credit and finance for agriculture, potentially undermining food production and raising food prices.[30] At the same time, the financial crisis may also have the effect of reducing government spending on social security, thus rendering the poor additionally vulnerable.

However, some measures taken to ameliorate climate change have had the unintended effect of increasing global food prices. In Chapter 4 it was suggested that bio-fuels may help in reducing Australia's carbon emissions in the transport sector, but only if certain problems are avoided. The diversion of crops and land into more profitable bio-fuel production may account for up to 70–75 per cent of the increase in international food commodity prices from 2002–08.[31] The 'first-generation' bio-fuels brought other environmental costs, such as deforestation and habitat loss, as well as controversy over their energy-intensive processing and the extent to which they actually reduce emissions.

Bio-fuels raised food prices by shifting crops from food to fuel, changing land use, encouraging commodity speculation, reducing grain stockpiles and inducing export bans in some countries.[32] The politics of bio-fuels also looked bad by allowing wealthy car owners in developed countries to keep driving at the expense of the human right to food in the developing world. The pursuit of 'second' and 'third' generation bio-fuels is mitigating the risk of food insecurity, by focusing on non-food sources of biofuel, such as grasses, and utilising more marginal, non-arable lands. Stronger international regulation of bio-fuel production is necessary to minimise its impacts on global food security.

The risk of food scarcity is also heightened if food security is dealt with as a *national security* problem rather than as a matter of *human security* affecting the world as a whole. The way to deal with declining global fish stocks, for instance, is not to assert unilateral 'resource nationalism', but rather to move towards more sustainable collective management of an increasingly scarce resource.[33]

In this regard, in response to higher food prices and food shortages, some countries (such as Qatar, Abu Dhabi, South Korea, China, India and South Africa) have begun to acquire access rights to vast tracts of agricultural land in developing countries, to secure stable, long-term food supplies for their own populations. Around 20 million hectares of farmland in Africa were leased under 40 to 99 year leases in 2008–09[34] in countries such as Kenya, Sudan, Ethiopia, Madagascar and Congo. On one hand, such schemes may provide much needed foreign capital to industrialise agriculture, and thus increase output, in underdeveloped countries where land is under-utilised, while simultaneously bringing revenue benefits for the developing countries.

On the other hand, modernisation brings serious risks if not managed properly. Local farmers or traditional graziers may be displaced and their livelihoods destroyed, without adequate compensation or alternative livelihoods being provided and with insufficient insensitivity to indigenous cultural attachments to traditional lands. Foreign agribusinesses may pursue unsustainable production methods which are ill-suited to local environments, while environmental law controls may be weak or under-enforced, destroying forests or other ecologically significant areas. Bio-fuel sources or export cash crops may become the focus instead of staple foods. A member of the Food and Agriculture Organization has warned that the leases adopted so far contain few procedural safeguards.[35]

Historically, during the colonial era, long-term concessionary contracts awarded to foreign investors to develop other essential resources, such as oil and gas, were frequently exploitative and prejudiced the ability of national governments to develop national resources in the interests of their own populations. There is certainly an irony in countries historically at risk of famine, such as Ethiopia, in signing over large areas of farmland to foreign investors to secure food supplies in wealthier countries. Leases without adequate protections have tended to stimulate conflict in various countries.

2 Historical links between climate change and conflict

There is some historical evidence to suggest that climate change tends to stimulate increased human conflict, particularly in connection with food production. One major study which analysed 8000 conflicts over 500 years, between 1400 and 1900, suggests that long-term climate change correlated with an increase in conflict and related population decline around the world.[36]

In particular, past periods of climatic *cooling* directly affected agricultural production (by shortening the growing season, restricting the availability of arable land and leading to crop failure), thus restricting food supply (and increasing food prices) and increasing the likelihood of conflicts, famines and epidemics. These factors ultimately reduced overall population size. The many other possible causes of war could not explain the temporal and spatial patterns of war shown by this research, which found that relative food scarcity is the key explanatory variable.[37]

While the historical patterns in that study related to periods of climate cooling, as noted earlier, today's higher temperatures (climate *warming*) will also adversely affect agricultural production. While warming may initially increase total bio-production, the negative effects of warming will outweigh the positive effects overall,[38] particularly when combined with the other impacts documented above (sea-level rises, ocean decline, disease, extreme weather, glacial retreat, energy price increases due to fossil fuel scarcity, resource disputes and social tensions). Continuing temperature acceleration 'might break the balance of human ecosystem that has been long established at a lower temperature'.[39]

Importantly, coping, adaptation and adjustment capacity can often defuse conflict. The study observes that the impact of climate change ultimately depends on the social mechanisms (including measures of adaptation and mitigation) deployed in response to it, which may include warfare, economic change, innovation, trade, migration, resource redistribution and the capacity of social and political institutions.[40] During the historical periods of climate cooling mentioned above, such responses often did not keep pace with ecological change, resulting in conflict and demographic decline.

Modern, industrialised agriculture may be less vulnerable to producing global famine, population decline and conflict than less sophisticated pre-industrial agriculture. Today there is also greater adaptive capacity than in the pre-modern world due to more sophisticated and robust social institutions for managing conflict (on the national and international planes), more advanced societies and better technology. At the same time, however, as already noted, the world now faces higher population stress, higher expectations about living standards and stricter political boundaries and border control (which limit migration choices and potentially fuel tensions).

In the long term, what are needed are more rational and equitable mechanisms for sharing scarce food resources across different countries. A good place to start is reform of the world trade system and the problem of agricultural protectionism. At present, import tariffs and export subsidies on agricultural products, particularly in the European Union and the United States, operate to prevent fair competition in agriculture by farmers in developing countries. To put it most starkly, an average European cow receives more per day in European Union subsidies (US$2.20) than the daily earnings of 1.2 billion people in poverty.[41]

3 The impact of agriculture and trade policy on food security

The impact of agricultural protectionism on global food security is complicated. On the one hand, import tariffs, export subsidies and market price support in developed countries tend to encourage agricultural output in those countries, which in turn increases global supply and lowers the price of agricultural commodities. Removing protectionism might therefore be expected to reduce agricultural profitability and output in developed countries, contracting global supply and raising food commodity prices in the short term. It might also reduce the food stockpiles available in developed countries for distribution as global emergency food aid.

On the other hand, liberalisation would enable farmers in developing countries to compete fairly in domestic markets of developed countries, probably resulting in increased agricultural output in most developing countries. Globally, this would counteract the decrease in food production in developed countries, shifting a greater global share of agricultural production to developing countries, as well as soon lowering any food price spikes which accompany liberalisation in developed countries.

As the Director-General of the World Trade Organization, Pascal Lamy, says: 'Through greater and fairer competition, international trade can help lower prices. But all of this presupposes that the trade-distorting agriculture subsidies that have given an unfair advantage to rich world farmers, will be tackled'.[42] Trade liberalisation is also expected to reduce the volatility of food commodity prices over time.

In addition, increased agricultural output in developing countries would increase incomes, employment, exports and GDP in developing countries as a whole. Even if global food prices did rise, so too would incomes in the developing world, such that food may still become more *affordable* even if it is more expensive. In China, for instance, trade liberalisation has tended to increase the wages of unskilled workers above inflation.

Removing all trade distortions would boost global welfare by US$300 billion by 2015[43] (about 63 per cent of which comes from agriculture). About 45 per cent of the global gain from liberalisation would go to developing countries, including an increase in the value of exports from developing countries of US$190 billion over 10 years (according to the World Bank). In developing countries, the poor would benefit, with full liberalisation reducing extreme poverty (less than US$1 per day) by 32 million people by 2015.

Liberalisation would inevitably create some 'losers'. Gains would not be spread evenly across all developing countries, nor within developing countries. Yet, since liberalisation will increase economic activity in most developing countries, it will also enhance the capacity of governments to provide food aid and social security to their citizens, as well as to help fund climate change adaptation strategies. Aid and development policies would still have an important role to play in assisting those whom the market fails, although reliance on aid is expected to decrease as liberalisation benefits developing economies.

Further, farmers in some developed countries lobby that their livelihoods would be destroyed by liberalisation. Yet, the reality is that the developed countries' share of global agricultural GDP would decrease only slightly from 30 per cent to 25 per cent by 2015. Even then, agriculture only accounts for a fraction of GDP in developed countries. A trade law expert at The University of Sydney, Brett Williams, says that 'removing a welfare diminishing policy will always make someone worse off, but keeping the policy will make even more people even more worse off'.[44]

Where removing protection makes uncompetitive farming unprofitable in developed countries, the appropriate policy response is to provide retraining and alternative employment opportunities, supplemented by social security – the same entitlement as any other failed business person. Agricultural protectionism currently operates as a proxy welfare system, at great cost to developed and developing countries alike.

Some developed countries may even gain from liberalisation. In Australia, many farmers welcome the opportunity to gain access to lucrative but restricted foreign export markets (such as the United States), since many Australian farming enterprises are efficient and competitive by world standards. Consumers in developed countries are also likely to gain, particularly where market price support measures – which raise domestic food prices – are eliminated.

More open agricultural markets are not the whole answer to the problems of food scarcity and unaffordability. Other necessary measures include stronger regulation of speculative commodity futures trading; capital investment in agriculture (including fertiliser, mechanisation, irrigation, transport infrastructure and sustainable development) and in research in developing countries;[45] and better sharing and improved transfer of agricultural technology and knowledge (including through common access to intellectual property rights in genetically modified organisms and gene technology).

In terms of emergency food aid for the poor, stronger global monitoring of food insecurity is needed, along with earlier intervention, the maintenance of adequate stocks of emergency food aid and enhanced capacity to rapidly distribute food aid. There has been increasing recognition of the 'right to food' within the United Nations system (including by the United Nations General Assembly in 2010) and African countries have insisted on a legal obligation on countries to cooperate and assist in the provision of food.

Yet, the reality is that many developed countries have tended to regard food aid as a discretionary political question of aid and development policy, rather than as a source of international legal obligation to the poor. The ongoing reform of the intergovernmental Committee on World Food Security is an opportunity to strengthen governance of food insecurity in light of climate change.[46]

C Water scarcity and conflict

Climate change is also making another essential resource, fresh water, increasingly scarce in places where it is already unavailable and much needed. The IPCC estimates that between 1.1 and 3.2 billion people could experience water scarcity by 2080.[47] In Asia, 60 per cent of the world's population currently enjoys only 36 per cent of the world's water resources.[48] Around 655 million people in Asia lack safe drinking water, while 1.9 billion people lack access to basic sanitation.[49]

The large number of people currently facing water scarcity or poor water quality may rise sharply as climate change affects surface water levels and curtails Himalayan glacial melt-water.[50] As the United Nations Secretary-General notes, high population growth, rising consumption, pollution and poor water management are contributing to a water crisis which is exacerbated by climate change: 'The consequences for humanity are grave. Water scarcity threatens economic and social gains and is a potent fuel for wars and armed conflict'.[51]

There are around 260 transboundary river systems in the world. Water scarcity is a potent source of potential human conflict, although human communities have often averted the risk of conflict through cooperative arrangements. Since at least the 19th century, countries which share freshwater resources such as rivers and lakes have developed sophisticated legal agreements to regulate the common resource. Many of the world's major transboundary river systems are governed by such legal regimes, which deal with issues such as the allocation of water rights between upstream and downstream users, as well as (less commonly) issues of water quality such as pollution. Where the total volume of water available is sufficient to meet the needs of all users, these arrangements have generally worked well.

It is when water becomes scarce that these arrangements are put to the test. Whilst outright 'water wars' are unlikely, there have long been unresolved tensions over the allocation of water in many international river systems, from the Nile to the Tigris and the Euphrates. Tensions are aggravated by increasing population pressure along many river systems and the decline in rainfall or glacial melt-water due to climate change.

1 Transboundary river tensions in Asia

Rivers systems on the Indian sub-continent provide good examples. In the Indus River valley, India enjoys a disproportionately larger share of water than Pakistan under the Indus Waters Treaty of 1960, which is likely to become an increasing source of tension as population and demand for water grows and as climate change decreases the water available.

Likewise, India's diversion of water from the Ganges River over the past 30 years has reduced the water available to its other neighbour, Bangladesh, resulting in increased salinity, a decline in river fish, droughts, declining sedimentary deposits, decreased agricultural productivity and loss of land to the sea.[52] As the flow of the Ganges decreases due to changes in rainfall, monsoons and glacial melt, population increases, and industry and agriculture uses more water, political tensions may increase.

So too is competition over the development of the Mekong River increasing among the Mekong countries of China, Vietnam, Thailand, Myanmar, Laos and Cambodia. A combination of uncoordinated flood mitigation measures, dams for industrial and hydroelectricity development, irrigation schemes and pollution is already affecting the availability and quality of water in the Mekong and impairing traditional livelihoods such as fishing. Climate change brings risks of further depletion of supply due to changing rainfall and monsoons and the increased salinity of delta river systems.

2 Water resources in the Middle East

Scarce water resources in the Middle East have been contentious for thousands of years, with a cultural history of mistrust and hostility in the region over the sharing of water resources.[53] In 1974, for example, Iraq threatened to attack a Syrian dam on the Euphrates River which diverted supply from Iraq. The flow of the Euphrates is projected to decrease by 30 per cent by 2100 as a result of declining rainfall associated with climate change, potentially straining relations between water users.

The Jordan River basin, shared between Israel, Palestine, Syria, Jordan and Lebanon, is a particular source of dispute and its flow is expected to decrease by up to 80 per cent by 2100. In 2001, for instance, Israel threatened armed intervention against Lebanon after Lebanon diverted water from a tributary river flowing into the Sea of Galilee, while Israeli-Syrian peace negotiations stalled in 2000 over access to Galilee waters. Israel estimates that its water supply could decrease by up to 60 per cent by 2100.[54]

Israel currently sources much of its water from the Occupied Palestinian territories. Israelis use four times as much water as Palestinians[55] and Israel severely restricts the capacity of Palestinians to develop their own water resources (such as by digging new wells). In Palestine violent protests have occurred over the resulting water scarcity and its economic effects (such as unemployment due to agricultural decline).[56] Many areas in the Middle East already face per capita water scarcity in the context of increasing contamination of supplies and limited opportunities to further utilise developed sources.

The lack of certainty in water-sharing arrangements often hinders development and generates further social instability. Increasing water scarcity may also complicate the prospects for permanent peace settlements in the Middle East and even unravel existing agreements.[57] The prospect of water scarcity may make governments less willing to negotiate. An 'island mentality' dominates government policy in the Middle East, where countries approach resources as a zero-sum game rather than cooperatively managing them.[58] Such an approach also carries over to the treatment of groundwater such as shared aquifers. The Disi Acquifer between Jordan and Saudi Arabia is being developed by Jordan (with French aid finance), with plans to extract unsustainable amounts of a non-recharging water resource. The aquifer is likely to be depleted within 50 years.

3 Internal social conflict over water

Conflict over water *within* countries is just as likely as inter-state conflict, particularly where the regulation of water resources between competing users is weak or ineffective, or where it disadvantages certain social groups (such as subsistence farmers or fishermen) over others (such as industry or agribusiness). In 2008, the Jordanian Army was used to quell robust protests against government attempts to control water theft in the Jordan valley.[59] Even in a developed country like Australia, there have been political and social tensions over the rapid decline of Australia's largest river system, the Murray-Darling, as drought has intensified competition for water among farmers, irrigators, industry, conservationists and urban populations.

In Latin America, while glacial melt-water is vital to the water supply of countries such as Peru and Colombia, over the past three decades the Andean glaciers have retreated by 25 per cent and may soon vanish.[60] Social conflict over scarce water, resulting from glacial melt and the drying of moorlands, has the potential to aggravate existing conflicts such as the rebel insurgency and resulting human displacement in Colombia.

In Peru, poor management of scarce water supplies has previously generated social conflict. In early 2004, large protests were held by the urban poor in three cities in La Libertad state against the privatisation of water supplies, which had left half of the state's residents without running water and contaminated irrigation channels with sewage.[61] The structural adjustment reforms demanded of Peru by the World Bank and the Inter-American Bank in the 1990s contributed to the collapse of water enterprises and resulting detrimental effects on the economy, public health and society. While the United Nations accepts that market principles have a role to play in providing water, the privatisation of water should not interfere with the basic human right to enjoy adequate water.

4 Cooperative water management defuses conflict

Conflict over access to water frequently arises because of a lack of mechanisms to enable participation by competing water users in decision-making about its use. In some regions, such as the Okavango River basin in Africa, arrangements for stakeholder dialogue and participation in Botswana and Namibia have helped to manage competing interests in the river system and defuse conflict before it arises. Likewise the Mekong River Commission has often functioned effectively to maintain relative harmony among competing water users, while there have been achievements through the Nile Basin Initiative and the Aral Sea Basin Commission.

In Europe, some countries have taken advantage of more formal judicial mechanisms to settle their water disputes peacefully, as in the 1990s when Hungary and Slovakia resolved a dispute over a dam on the Danube River in the International Court of Justice. Some have suggested creating a United Nations arbitration mechanism for the resolution of transboundary water disputes.[62] Many transboundary water governance arrangements tend, however, to be voluntary or consensual rather than more strictly legalistic.

Most countries remain reluctant to submit water disputes to binding adjudication and prefer to utilise more informal methods such as negotiation, diplomacy or river commissions over which they retain more control. There is no 'one-size fits all' model that is suitable for all river systems. There are, however, a range of models available which countries can build upon as 'best practices' to meet the challenges of increasing water scarcity resulting from climate change.

Emerging legal regimes governing transboundary watercourses and groundwater acquifers will need to factor climate change impacts into the implementation of key norms such as the principle of equitable and reasonable utilisation of water, the obligation not to cause significant harm and to adopt a 'precautionary approach' to safely using resources, and in the protection and preservation of ecosystems.[63]

Ultimately, whatever the shape or form of regulation, there must be a shift away from the old view in international relations that any country may exercise exclusive and permanent sovereignty over its own natural resources.[64] In a world facing resource scarcity, instead what is needed is stronger multilateral cooperation for a more rational and equitable distribution of resources that are essential to human dignity.

D Interaction with social, political and economic instability

The direct impacts of climate change described above are likely to aggravate existing structural weaknesses in some countries. The effects of climate change may aggravate existing economic weaknesses, such that unemployment grows, the authorities cannot meet the basic needs of society,[65] and internal conflict over the distribution of scarce resources follows. Further, political instability and poor or weak governance make it harder for a country to adapt to the effects of climate change and to resolve conflicts without violence.[66] As a result, conflict over essential but increasingly scarce resources – water, food, land and energy – may increasingly arise, and the political legitimacy of the national authorities is increasingly called into question.

Already some resource conflicts over land have been prompted in part by the interaction of climate change related factors (such as desertification) and the incapacity of political institutions to manage problems of resource scarcity and distribution. The impoverishment of nomadic peoples due to drought in Mali in the 1970s, coupled with the political and social marginalisation of nomads and their resource claims, was one factor contributing to the 'Second Tuareg Rebellion' in 1990.[67] Likewise tensions over desertification, resource scarcity and political aspirations have affected northern Nigeria, Kenya and Ghana.

Even in the most developed countries, a lack of effective government responses to the impacts of climate change can also produce internal social unrest. The inadequate disaster management of the United States in the aftermath of Hurricane Katrina in 2005 is a case in point. While the southern United States has long been battered by hurricanes, there is emerging evidence that global warming is intensifying their destructiveness.[68] Katrina was among the most intense hurricanes ever recorded there. A disintegration of law and order followed in the wake of the hurricane and the United States eventually deployed more than 50,000 police, military and paramilitary personnel to restore order. A high level of social, ethnic and political conflict resulted from the inability of the United States authorities effectively to manage expectations and to stabilise the situation.

One facet of climate change which will place pressure on weak governments is its public health impacts. It is self-evident that a scarcity of food and water will have negative repercussions for human health and human security, as will the more direct impacts of climate change such as natural disasters (which take lives and destroy shelter and livelihoods). According to the World Health Organization (WHO), 'climate change poses substantial risks to human health, particularly among the poorest populations'.[69] The WHO estimates that the effects of climate change from the mid-1970s to 2000 caused at least 150,000 deaths.[70] Climate change affects the conditions necessary for good public health.

Specific threats to health include increasing heat waves, rainfall changes affecting water supplies, rising temperatures and the impact on agriculture, rising sea-levels and coastal flooding, and population displacement.[71] Certain diseases (such as cholera, typhoid, hepatitis, giardia, malaria, dengue, plague and encephalitis) may spread to new areas and persist for longer, and affect populations which do not have immunity or lack health resources.[72] Further, climate-induced migration may shift disease patterns; crop failure and famine may reduce resistance to infection; water scarcity and contamination may increase disease transmission; and health infrastructure will come under strain.[73]

Already most developing countries are unlikely to reach the health targets of the Millennium Development Goals by 2015 and the effects of climate change are likely to aggravate inequities in health between developed and developing countries.[74] As governments appear less able to protect public health, so do they come to be seen as less legitimate and more vulnerable to radical political challenge and change.

E Human displacement and social conflicts

Displacement related to climate change, discussed in the previous chapter, may also aggravate social tensions and other conflict. There is scant evidence that newly displaced persons are likely to start 'climate wars' in host territories. It is important not to simplistically depict migrants as violent people itching to create conflict. The spectre of 'migratory warfare'[75] is part of the broader, and often alarmist, security discourse which warns of 'climate wars'.

More often, risks of conflict arise from how (often unwelcome) forced migrants are received by others, including host communities and local governments. Political sensitivities about migration control, inflammatory ethno-religious or political agendas and isolationist politics may heighten the risk of confrontation.[76]

A case study of Bangladesh provides a useful illustration of the subtle processes by which social tensions and other forms of conflict can sometimes be associated with migration. Bangladesh is a good example because it is often presented as one of the 'worst case scenarios' due to the severe impacts of climate change there, the large numbers of people thought to be at risk of displacement, and the complex regional security situation.

1 Social conflict over scarce resources

As noted in the previous chapter, if most climate change related displacement will be internal, it makes sense to first focus on any risks of social strife or other tensions which might arise within Bangladesh. The internal displacement of people there as a result of the slow and rapid onset climate change processes described in the previous chapter will result in increased competition over key resources such as land, housing, food, water and employment. Already in Bangladesh, coastal disasters, such as flooding, storms and cyclones and rising sea-levels, have resulted in substantial migration away from coastal areas. Resulting social tension has already taken a number of forms in Bangladesh.

First, intense competition over 'free' but scarce public *khas* land has led to tension, violence and insecurity between migrants themselves, as well as between migrants and local communities.[77] Secondly, tensions have arisen where private landowners have evicted migrants from their lands, resulting in reprisal violence by migrants.[78] Thirdly, in some forest areas, conflict has arisen between the forest dwellers and migrants from the plains over the utilisation of forest resources.[79]

Fourthly, large-scale, unplanned rural migration to urban slums as a result of climate processes has generated social tensions as well as human insecurity for slum-dwellers. Slum residents are highly vulnerable to forced evictions, which have provoked unrest and violence between residents, police and government authorities. Further, slum residents are stigmatised as bringing crime and violence, even though they themselves are extremely likely to be victims of serious crime and lack protection from the police.[80]

The entry of new slum residents into the labour market also brings real and perceived competition for employment with local residents, particularly in the informal sector (such in rickshaw or van pulling, bus or truck driving, domestic work and cleaning, and in the garment industry).[81] The abundant supply of labour can also lead to lower wages, which in some instances has provoked attacks on migrants.[82]

Even if the economic position of migrants improves by moving to the cities, they often become more socially marginalised, cut off from their traditional community and support networks in rural areas, and face social and security concerns.[83] Slum-dwellers typically experience social exclusion from basic services and infrastructure, including adequate housing, clean water, sanitation, healthcare, education and electricity.[84] Government housing and social support for slum residents has also been limited, while resettlement and rehabilitation of evicted slum residents is seriously inadequate.[85]

Finally, slum residents are also insecure because they are at risk of multiple displacement due to the vulnerability of slum areas to climate change. Dhaka, in particular, is low-lying (2–13 metres above sea-level) and even moderate sea-level rise may inundate substantial parts of it.[86] Slum residents tend to live in low-lying, flood-prone, water-logged and poorly drained areas with inadequate sanitation.[87] They are likely to be particularly affected by the intensification of flooding, aggravated by the reality that one third of slum houses (typically of tin and bamboo) are too fragile to resist disasters.[88]

It is also worth noting that previous government efforts to relocate large numbers of people within Bangladesh caused serious violence. Between 1973 and 1997, up to 600,000 people migrated internally to the Chittagong Hill Tracts, resulting in deforestation and erosion, and contributing to conflict with indigenous peoples in the area. That experience is a lesson in how not to manage internal resettlement initiatives, which will be increasingly needed to assist those displaced by climate processes.

2 Risks of radicalisation and extremism in Bangladesh

It has been suggested that the impacts of climate change on Bangladesh may bring a second form of conflict – an increase in Islamist extremism or terrorism.[89] As climate change disrupts livelihoods and displaces people from their local community, so it is thought that the resulting impoverishment, frustration, social disconnection, alienation and disempowerment may attract some people to radical groups.[90] Such groups are attractive because they provide charitable and social services to the poor (where government assistance is very limited),[91] educational opportunities in religious *madrasas*, and a sense of certainty, purpose and identity for the displaced.

Such a process cannot be ruled out in exceptional cases, although it is notable that some senior Islamist terrorists in Bangladesh have come from the educated middle class rather than the poor and disempowered. It is important not to exaggerate any risks of radicalisation in Bangladesh, which one senior Bangladeshi military official describes as 'unlikely' and 'unrealistic'.[92] Religious extremism is largely alien to Bangladeshi society and culture, and is strongly opposed by Bangladesh's vibrant civil society, the media, and many religious and political leaders. The use of suicide bombing by extremist groups some years ago provoked a popular backlash and tougher government law enforcement.[93]

3 Trans-border conflict with India

It is widely claimed that climate change will stimulate large-scale cross-border emigration from Bangladesh and thus generate a third kind of conflict - transnational security threats.[94] As noted earlier, however, much climate-related movement in Bangladesh will be internal rather than cross-border; most longer term movement will be gradual not sudden; that which is sudden will often involve temporary relief rather than permanent migration; and those who are able to move internationally will tend to be wealthier rather than poorer. These realities of movement dampen many of the bullish predictions about the likely security implications of displacement.

It remains true that some cross-border migration may result from the impacts of climate change, particularly if greater internal displacement and the resulting urbanisation are not well managed by Bangladesh. Some tentative conclusions can be drawn about its likely security consequences. At a minimum, war with India is highly improbable. As a senior Bangladeshi military official said, 'two democracies do not fight',[95] and the migration of civilians across borders alone is hardly incendiary enough to trigger armed conflict.

Even so, tensions and violence on the border over the movement of people have been all too common. In recent decades, India has 'securitised' its border with Bangladesh by building a border fence and deploying 45,000 Indian Border Security Force (BSF) personnel to patrol it.[96] It is estimated that 843 Bangladeshi nationals were killed at the border by the BSF between 2000 and early 2010.[97] There have also been occasional violent clashes directly between Indian and Bangladesh border forces.[98]

This legacy of border security suggests that sustained climate-related movement from Bangladesh into India may stimulate further tensions and loss of life. Much will depend on the scale of movement, the destinations within India, the manner in which migrants are received by host communities and the political response to migration by India.

As one Bangladeshi non-governmental organisation notes, migration only becomes a security problem when it is politicised.[99] Traditional 'innocent', but unauthorised, cross-border movements to India – by farmers, traders, graziers and so on – are now securitised as a result of concerns in India about the infiltration of insurgents, terrorists and illegal labourers.[100] The inflammatory politics of ethno-religious political parties in India (such as Hindu nationalists) have even portrayed Bangladeshi migrants as terrorist threats.[101]

The most sensitive and complicated area for potential tensions from cross-border migration concerns India's north-eastern states and their borders with Bangladesh. More than 50,000 deaths have occurred since India's independence in 1947 as a result of ethnic, tribal, secessionist, insurgent or terrorist violence in the seven north-eastern Indian states,[102] with up to 120 insurgent groups based there.[103]

Migration from Bangladesh has caused friction in some of these conflicts. For example, India has claimed that up to 10 million Muslim Bangladeshis have migrated to Assam,[104] but those figures are unverified. Assamese groups have militantly opposed large-scale 'illegal migration' from Bangladesh on the basis that migrants have altered the demographic balance and thus the distribution of political power, dominated employment and the exploitation of natural resources (such as oil, timber, tea and water)[105] and dispossessed tribal people from their lands.[106]

It is, however, too simple to 'blame' illegal migrants for such conflicts in India, which again are a result of a range of complex factors. These include a failure by the Indian central government to fund development in remote areas and the unfair exploitation of natural resources;[107] conflicts between competing Indian groups, and between different indigenous groups; populist 'anti-foreigner' movements which exploit migration for political advantage; and even criminal extortion.

Ultimately the case study of Bangladesh challenges many commonly held assumptions about the links between climate-related displacement and conflict, and illustrates the complexity of that relationship. While Bangladesh is often presented as a good example of the risks of 'climate conflict', in fact the above analysis of previous migration to India suggests that: war between governments is unlikely; migrants do not 'cause' conflicts in India, but are often exploited by others seeking to use violence for different purposes; the risk of Islamist radicalisation in Bangladesh is relatively small; and migrants are typically more threatened (as with the risks faced at the Indian border) than threatening.

F Resource conflicts: the Polar regions

In addition (and sometimes connected) to conflicts animated by resource scarcity or displacement, further (and more conventional) conflicts over territorial and maritime boundary disputes may arise as a result of climate change. In the Arctic, for example, the rapid melting of polar ice – with predictions of an ice-free Arctic by 2060[108] – opens up the region to newly accessible, commercially exploitable maritime trade routes, as well as resource exploitation on the seabed.[109] Such developments may aggravate long-standing maritime disagreements between, for instance, Canada and the United States over the status of certain Arctic waters as international straights (open to 'innocent passage' by all countries' ships) or internal waters (only open to shipping authorised by Canada).

As regards the Arctic seabed, a number of states assert competing claims over their continental shelves (the sea floor naturally connected to the geological plate on which a country's land rests). Under the international law of the sea, recognition of a country's claim to its continental shelf carries with it rights to exploit resources found there. Russia has duly asserted continental shelf claims under the Arctic, exemplified by its infamous planting of a Russian flag on the seabed by submarine in 2007. Denmark has also staked Arctic claims, while the United States is moving in the same direction. At the same time, the area is becoming increasingly militarised, with Canada announcing a $5 billion Arctic defence plan in 2008 in order to protect its sovereign rights from unwanted competition.

Whether competition over the unlocking of Arctic resources escalates into military confrontation is an unknown possibility. Despite the increasingly militant posturing by some Arctic nations, there are also promising signs that competition over resources may be handled within the existing framework of the international rule of law. Paradoxically, the consequences of climate change in the Arctic, along with rising energy costs, led former United States President George W Bush to signal that the United States was finally contemplating joining the United Nations Convention on the Law of the Sea of 1982,[110] which is the comprehensive global framework governing maritime claims and resources.

While the Republicans historically opposed that treaty in the United States Senate, the United States has realised that becoming a party to that treaty provides the only real avenue through which the United States can claim and enforce any sovereign rights to exploit resources in its Arctic continental shelf, particularly when faced with competing legal claims in the area. The treaty contains technical rules on the measurement and allocation of the continental shelf, empowers countries to exploit their resource rights and provides independent mechanisms to resolve disputes arising between countries. While some areas could benefit from clarification in relation to the Arctic,[111] the general framework of law may be adequate to accommodate and peacefully reconcile the competing interests in the area.

The international legal regime governing Antarctica is rather different and in some ways more vulnerable. Up to the late 1950s, a number of countries sought to claim parts of Antarctica as their own sovereign territory. However, many countries also recognised the special scientific and ecological importance of Antarctica, leading to the adoption of the Antarctic Treaty of 1959. Under that treaty, any national claims to sovereignty (that is, exclusive ownership) over territory in Antarctica - including Australia's claim of approximately half of the continent - were suspended. In that sense Antarctica does not 'belong' to any particular countries and its natural resources cannot be freely exploited and disposed of in the same way as resources in a country's own territory. Subsequent legal agreements have prohibited commercial mining and ensured the protection of the Antarctic environment.

As resources becoming increasingly scarce (whether oil, gas or fisheries) and as mining technology becomes more cost-effective in remote areas, major powers may be increasingly tempted to exploit the rich resources of Antarctica and its marine areas. China, for instance, is substantially increasing its presence in Antarctica. As competition for Antarctic resources grows, there may be increasing pressure to unwind the special protections of the Antarctic Treaty System. In particular, suspended territorial claims might be actively asserted; other countries might also seek to claim territory; exploitation of resources might begin in earnest; and countries may begin to withdraw from the Antarctic Treaty System – without replacing it with anything as protective. The collapse of the special regime of Antarctic governance, and resource competition, bring with it risks of tension and conflict between countries with different interests.

G Resource conflicts: maritime boundary disputes

It is not only the polar maritime areas that carry risks of conflict associated with climate change. Maritime boundaries elsewhere are also likely to be affected. Sea-level rises (or changes in shared boundary river courses, lakes or wetlands, driven by climate change) place pressure on the maintenance and stability of existing maritime and territorial boundaries and claims. On one estimate, only 160 of 365 potential maritime boundaries are already determined,[112] and the shifting of boundaries through rising sea-levels may complicate the settlement of already difficult disputes. Maritime boundary disputes have often triggered confrontations between navies, as in the South China Sea, where numerous countries claim resource-rich areas around the Spratley Islands.

In the Pacific Ocean and South Asia, the likely loss of land territory due to rising sea-levels may lead to the disappearance of islands or other maritime formations, or result in certain areas becoming uninhabitable. Drought, salination, storms and sea-level rises are seriously affecting island states such as Micronesia, Tuvalu and Kiribati, while in the Maldives a two-metre sea-level rise would inundate 1200 islands.[113] At the same time, the coping capacity of some of these states is being decimated by other economic impacts of climate change; coral bleaching in the Pacific, for instance, may result in up to a 50 per cent loss in GDP in Pacific Islands by 2020,[114] while overfishing by Asian countries is further reducing the ability of island countries to finance their own adaptation.

In low-lying Bangladesh, predicted sea-level rises are likely to alter dramatically the coastline, deltas and river boundaries.

Already Bangladesh is locked in disputes with neighbouring India and Myanmar over its maritime boundaries, which have occasionally resulted in naval incursions into Bangladeshi maritime areas.[115] The retreat of its coastline could prejudice its claim to maritime areas containing rich hydrocarbon resources, further impoverishing a highly underdeveloped country.

Currently, the international law of the sea allows countries to assert certain rights and economic interests in designated sea areas extending from their coastlines. Thus the territorial sea extends 12 nautical miles out to sea from the coastline and countries may exercise full sovereign rights in that area. A 'contiguous zone' extends to 24 nautical miles from the coast, in which countries can regulate a more limited range of matters such as customs, immigration, pollution and fiscal issues. A country can also claim a 200 nautical mile 'Exclusive Economic Zone' in which the country enjoys rights to exploit the economic resources found within that water area, while rights can also be asserted over certain resources found in or on the seabed within a country's continental shelf.

Maritime boundaries are generally measured from the low-water mark on the coastal landmass, although the measurements become more complicated when dealing with indented coastlines (such as bays) or in countries with large numbers of islands (such as archipelagic countries like Indonesia). As sea-levels rise with climate change, coastlines may move inwards and with them the potential for maritime boundaries – and rights to the economic resources which depend on them – to change accordingly. At present, the law of the sea does not envisage any rules or procedures for modifying maritime boundaries and zones as coastlines shift away from previously established positions.

Recently, the Commission of the European Union has enigmatically suggested that '[t]here might be a need to revisit existing rules … as regards the resolution of territorial and border disputes'[116] – but without indicating how the rules should be revised. Some have suggested that 'permanently fixing ocean boundaries' may be preferable to the increasing uncertainty (and potential for conflict), following sea-level rises, generated by the existing 'ambulatory' (or wandering) nature of baselines and the maritime boundaries and resource rights which flow from them.[117] Such a suggestion could be important in areas where sea-level rises spell the end of permanent habitation and thus the ability to claim maritime areas and attendant economic rights could otherwise disappear. At any rate, some clarification or development of the law of the sea may be necessary to ensure that countries enjoy more certainty about their maritime and resource rights as seas rise, so that the potential for conflict is defused at any early stage.

H Stronger global governance of climate change and its security risks

Averting the complexity of compounding conflict risks associated with climate change requires a system-wide, preventive and proactive approach from a range of organisations.[118] Stronger collective management of the causes of climate-related conflict is necessary to prevent the risk of conflict from materialising, including more equitable management of food scarcity and water resources, and reform of regulatory approaches to world trade and so on.

1 The militarisation of climate change

As mentioned earlier, some national militaries have begun turning their attention to the national security risks of climate change. This conventional 'securitisation'[119] of climate change risks interfering with more cooperative, multilateral efforts to deal with the problem. Unilateral national military responses to the problems of climate change may be limited to addressing the symptoms (such as food riots or social unrest) rather than the causes (such as agricultural protectionism, carbon emissions and so on).

National military responses may also escalate tensions by pursuing competitive 'resource nationalism' – including securing access to resources by force – rather than cooperative, peaceful, negotiated and fair distributions of scarce resources. Since a military response is a form of exceptional or 'emergency' action, it will typically not involve participatory, deliberative dialogue about how to best resolve social unrest or violent tensions – and which is usually necessary in guaranteeing lasting solutions.

That is not to suggest that national militaries should have no role at all in responding to climate change risks. Given their resources and rapid response capabilities, national militaries can assist countries at risk in a range of humanitarian, peace-building and stabilisation functions.[120] The United Kingdom Ministry of Defence, for instance, has partnered with British meteorologists to study countries likely to be affected by climate change and food and water scarcity, in order to identify where United Kingdom military assistance may be needed.

During the 1990s and 2000s, the Australian Defence Force and law enforcement agencies significantly expanded their peace-building and stabilisation capacities, with missions in troubled countries such as East Timor and the Solomon Islands, and greater experiences of humanitarian and disaster relief after the Indian Ocean tsunami. Australia is well positioned to use its military and other authorities to assist in averting both the human and national security threats which are linked to climate change.

What matters is that national militaries are deployed with the consent of countries being assisted, or are otherwise coordinated and authorised by the collective security mechanisms of the United Nations system. The security architecture of the international system certainly needs to become more responsive to the risks of climate-related insecurity, not least so as to defuse calls for risky unilateral action.

2 Collective and multilateral responses

Engaging preventive diplomacy at an early phase can assist in dealing with the root causes of climate-related insecurity and related causes (such as inequality, weak governance and so on), moving beyond the traditionally more narrow focus of diplomacy on averting imminent physical risks of violence. The European Commission, for example, has committed itself to developing new analysis, monitoring and early warning capabilities to avert conflict risks associated with climate change.

At the global level, the United Nations Security Council has already turned its attention to the conflict risks associated with climate change, as noted earlier. The Council also signalled its willingness, in Resolution 1625 (2005), to 'adopt a broad strategy of conflict prevention, which addresses the root causes of armed conflict'. Despite the Council's increasing concern about 'human security' or non-traditional security threats, including climate change, the international community remains divided on whether the Council is the appropriate forum for addressing such matters.

Many western states and Pacific Island nations favour Council action. The former are in favour because they have more control over the decisions of the Security Council (given its limited and non-representative membership) than over the General Assembly (which has a universal membership in which western and developed states are outnumbered). The latter are in favour because 'securitising' climate change brings greater international attention and visibility to the issue.

In contrast, the Non-Aligned Movement, the Group of 77, and major powers such as China and India are unconvinced that the Security Council should play a role, partly because they believe that other fora – such the General Assembly, ECOSOC, the Commission on Sustainable Development or new specialised bodies – are better placed to respond.[121] Even when the issue of security and climate change was debated in the General Assembly in 2008–09, the resolution ultimately adopted was a much watered-down version of an original draft which had referred to the power of the Security Council to identify and respond to threats to international peace and security.[122]

There is likely to be most consensus around Council action in situations where conflicts have already crystallised and become acute – rather than at an earlier preventive stage – by which time it may be too late to avert humanitarian crises (particularly where other organs have failed to take preventive action). Given that national militaries are increasingly concerned about the security risks posed by climate change – leading to national competition to secure access to dwindling global resources – it is imperative that the Security Council exercise collective control over the human security problems presented by climate change, so as to manage and defuse sovereign competition over dwindling resources. Using the Council is also attractive because it is the only international body with the legal power (under the United Nations Charter) to bind all countries to take firm action against a threat to security.

Regional organisations too must be more active in collectively managing the effects of climate change and resource scarcity in their own geographical regions. There are already some tentative developments: the Asian Development Bank plans to double investment in Asian water projects to $2 billion per year, given the potential for conflict if water governance remains weak;[123] the European Commission is developing its own Arctic Policy, in part to avert conflict risks; and there is debate over whether NATO should shift further away from conventional security towards concerns about 'energy security'.

Institutions involved in aid and development projects, such as the World Bank, International Monetary Fund, United Nations Development Program and donor agencies, should also address climate-related security risks. The World Bank, for instance, estimates that 40 per cent of its development projects are affected by climate change,[124] yet the Bank has been increasing its lending for fossil fuel energy relative to renewable energies.[125] In environmental assessments, the effects of climate change need to be considered in judgments about the sustainability of development projects, so as to avert potential social or other conflicts arising over unsustainable or ill-devised projects. Where nuclear energy is financed, it must be ensured that sufficiently stringent safeguards are applied to the protection of nuclear material, so that new nuclear security risks are not created.

Climate change impacts should also be factored into the approvals process for international loans and assistance, and multilateral or bilateral foreign aid, while climate change adaptation and mitigation measures could be made conditions of finance or assistance. At the same time, such conditionality must be sensitive to the risk of forcing premature energy reductions onto developing States, which itself risks generating conflict if developing states feel aggrieved that their development is being thwarted by already developed donor countries. In international investment law, there are various impediments to the implementation of climate change mitigation measures[126] which require attention if investment is not to aggravate conflict risks.

3 Development aid and humanitarian relief

Finally, humanitarian organisations are likely to find themselves increasingly stretched in responding to the range of disasters and emergencies brought about by climate change (such as floods, droughts, heat-waves, fires, storms and landslides). Oxfam International estimates that the number of people affected by climate-related disasters doubled in the past 30 years (from 121 million to 243 million people per year) and is likely to increase by a further 54 per cent to 375 million people per year by 2015.[127] The effects of such emergencies are not merely temporary, but can have long-term impacts on livelihoods, housing, health and security. Oxfam believes that a doubling of global funding for emergency aid is needed to deal with the growing risks.

Australia's aid policies and its development agency, AusAID, may help to mitigate the security risks of climate change in a number of ways. From 2008–09, Australia has committed $464 million to a global food security initiative, aimed at enhancing agricultural productivity, improving rural livelihoods and building community resilience. During 2008–11, Australia increased its development assistance in the area of water and sanitation by $300 million, including 'water security' initiatives to better manage freshwater resources in the Asia-Pacific, building on Australia's existing technical assistance to bodies such as the Mekong River Commission.

Australia also committed $150 million in 2008–11 to climate change adaptation initiatives in the Asia-Pacific region, as well as supporting efforts at Copenhagen to increase and enhance global climate change finance. Given the increase in extreme weather events related to climate change, Australia's adoption of a disaster risk reduction policy may assist in reducing the likelihood of disasters (through flood protection mechanisms, livelihood diversification and safe building practices) and in strengthening the capacity of communities to respond and cope with disasters.

These initiatives are additional to Australia's long-term efforts in improving governance in weak countries (by strengthening law and justice, civil society, democracy, the public sector and economic management) and regional stability (by tackling transboundary threats such as pandemics, disasters, climate change and transnational crimes). Preventive activities not only minimise the human hardship which otherwise ensures from reacting too late, but are also typically cheaper than waiting for conflict to crystallise.

To summarise, the following table sets out some of the ways in which the international community can prevent and mitigate the risks of climate-related conflict:

Develop Legal Standards	Enhance Institutions and Governance
• Reduce carbon emissions • Strengthen adaptation frameworks and funding • Prevent food crop diversion into bio-fuels and limit commodities speculation • Limit exploitative foreign land concessions • Pursue trade liberalisation in agriculture • Commit to binding treaty norms on transboundary water governance • Agree principles for resolving Arctic maritime disputes and other ambiguities in the law of the sea • 'Mainstream' climate change issues in developmental processes	• Improve global food security, including emergency preparedness and capability • Safeguard human rights to water from commercial exploitation • Strengthen conflict early warning, prevention and diplomacy • Enhance mechanisms for binding dispute settlement • Enhance global public health cooperation • Increase technology transfer to developing countries • De-escalate border tensions • Resolve maritime boundary disputes • Encourage collective security responses and resist militant unilateralism

I Conclusion

While the spectre of 'climate wars' is fanciful, the effects of climate change will stimulate domestic and global tensions and aggravate existing causes of social, political and military conflict. The effects of climate change are more likely to contribute to 'human insecurity' rather than conventional wars, in particular by worsening food insecurity, fresh water scarcity, land shortages and public health. In some cases, local resource scarcities may stimulate social conflict between different groups competing for land, water or food, including where displacement and migration increases pressures.

Such consequences do not inexorably lead to outbreaks of violence, political instability or social unrest, but they may be contributing factors given the right constellation of conditions (such as weak economies or governments, or divided social orders). In this way, climate change multiplies the risk of conflict rather than simply 'causing' it. At the national level, countries may also find themselves in disputes about rights to exploit maritime resources in areas where sovereign boundaries are unsettled, or to use resources in increasingly accessible areas like the polar regions. If not carefully managed, such disputes about sovereign rights have a historical tendency to escalate towards violence.

With the processes of climate change already underway, even the rapid conclusion of an effective post-Kyoto global climate change regime would not be sufficient to avert at least some of the security risks that are already underway. It is vital, therefore, to think about how the international legal system as a whole can be utilised and reformed to prevent, contain and remedy the possible security threats. More cooperative multilateral global governance and institutions, more equitable principles for the sharing and distribution of scarce resources, and stronger mechanisms for preventing and defusing conflicts as they arise are all good places to start.

Notes

1 Mabey, N, 2007, 'Delivering Climate Security: International Security Responses to a Climate Changed World' 69 *Whitehall Papers* 1 at 2.

2 Smith, D and Vivekananda, J, 2007, *A Climate of Conflict: The Links between Climate Change, Peace and War* International Alert, London, available at <http://www.international-alert.org/publications/pub.php?p=322>.

3 UN Security Council, 2007, 'Security Council Holds First-Ever Debate on Impact of Climate Change on Peace, Security, Hearing over 50 Speakers' UN Doc SC/9000, UN Department of Public Information.

4 European Council, 2008, 'Paper from the High Representative and the European Commission to the European Council, Climate Change and International Security' Doc S113/08, p 6.

5 UN General Assembly, 2009, 'Climate Change and its Possible Security Implications' UN Doc A/RES/63/281.

6 Townsend, M and Harris, P, 22 February 2004, 'Now the Pentagon Tells Bush: Climate Change Will Destroy Us', *The Observer*, London; Australian Defence Force (ADF), 2007, *Joint Operations for the 21st Century*, ADF, Canberra; Allard, T, 17 May 2007, 'Climate Threat in Military's Sights', *Sydney Morning Herald*, Sydney; ADF, 2007, *Climate Change, The Environment, Resources and Conflict* reported in Pearlman, J and Cubby, B, 7 January 2009, 'Defence Warns of Climate Conflict', *Sydney Morning Herald*, Sydney; Campbell, K et al, 2007, *The Age of Consequences: The Foreign Policy and National Security Implications of Global Climate Change*, Center for Strategic and International Studies, Center for a New American Security, Washington DC; Dupont, A and Pearman, G, 2006, *Heating Up the Planet: Climate Change and Security*, Lowy Institute Paper 12, Sydney.

7 A term used by Dyer, G, 2008, *Climate Wars* Scribe, Melbourne. For a clear analysis of how this discourse has been perpetuated, see Elliott, L, 'Climate Migration and Climate Migrants: What Threat, Whose Security?' in McAdam, J (ed), 2010, *Climate Change and Displacement: Multidisciplinary Perspectives*, Hart Publishing, Oxford, p 175.

8 See, eg Barnett, J, October 2001, 'Security and Climate Change', Tyndall Centre for Climate Change Research Working Paper No 7, p 5; Barnett, J and Adger, N, 2005, *Security and Climate Change: Towards an Improved Understanding*, Paper presented at the Human Security and Climate Change International Workshop, Asker, p 11.

9 Dietz, S, 2007, 'The Impact of Climate Change on International Economic Relations: Perspectives from the Stern Review' 13 *The Brown Journal of World Affairs* 173 at 178; Podesta, J and Ogden, P, 2008, 'The Security Implications of Climate Change' 31 *The Washington Quarterly* 115 at 119.

10 See, eg, Smith and Vivekananda, above n 2, p 12; see also Edwards, S, 2008, 'Social Breakdown in Darfur' 31 *Forced Migration Review* 23 at 23; Integrated Regional Information Networks (IRIN), 22 May 2008, 'Sudan: Rich Farms, Conflict and Climate Change' UN Office for the Coordination of Humanitarian Affairs, Nairobi.

11 European Council, above n 4.

12 Smith and Vivekananda, above n 2.

13 See, eg, Greek Chairmanship of the Human Security Network, May 2008, *International Conference on Climate Change and Human Security*, Athens; Wisner, B et al, 2007, *Climate Change and Human Security*, Peace Research and European Security Studies, Mosbach; International Institute for Strategic Studies, 2007, *Strategic Survey 2007*, London.

14 Smith and Vivekananda, above n 2, p 4.

15 Adam, D, 9 April 2008, 'Food Prices Threaten Global Security – UN', *The Guardian*, London.

16 Brown, LR, 2009, *Plan B 4.0: Mobilizing to Save Civilization*, Earth Policy Institute, Washington DC.

17 McKie, R, 4 October 2009, 'Arctic Seas Turn to Acid, Putting Vital Food Chain at Risk', *The Observer*, London, p 16.

18 Ibid.

19 Epstein, R, 30 April 2008, 'UN Chief Warns of Civil Unrest Amid World Food Shortage', *ABC News*, available at <http://www.abc.net.au/news/stories/2008/04/30/2231223.htm>.

20 Vidal, J, 17 April 2008, 'Nations Split on Ways to Tackle Hunger Alert', *Sydney Morning Herald*, Sydney.

21 US National Intelligence Council, 2008, *Global Trends 2025: A Transformed World*, USGPO, Washington DC, p 51.

22 Paskal, C, 2007, *How Climate Change is Pushing the Boundaries of Security and Foreign Police*, Energy, Environment and Development Programme, EEDP CC BP 07/01, Chatham House, London, p 8.

23 McKie, R, 13 December 2009, 'Why Britain Faces a Bleak Future of Food Shortages', *The Observer*, London.

24 International Food Policy Research Institute (IFPRI) for the World Bank and Asian Development Bank in Goldenberg, S, 30 September 2009, 'By 2050, 25m More Children will Go Hungry as Climate Change Leads to Food Crisis', *The Guardian*, London.

25 Brown, O and Crawford, C, 2009, *Rising Temperatures, Rising Tensions: Climate Change and the Risk of Violent Conflict in the Middle East*, International Institute for Sustainable Development, Winnipeg.

26 Smith and Vivekananda, above n 2, p 11.

27 Ibid.

28 Walt, V, 27 February 2008, 'The World's Growing Food-Price Crisis', *Time Magazine*.

29 Mitchell, D, 2008, *A Note on Rising Food Prices*, World Bank Policy Research Working Paper, Washington DC, pp 16–17.

30 Food and Agricultural Organization of the United Nations (FAO), 2008, *Food Outlook*, FAO, Rome.

31 Mitchell, above n 29, pp 16–17.

32 Ibid.

33 See, eg, Stephens, T, 2008, 'Fishing-Led Development in the South Pacific: Charting a Pacific Way to a Sustainable Future' 39 *Ocean Development and International Law* 257 at 257–86.

34 28 November 2009, 'Ethiopia Being "Used like an Empty Womb"', *Sydney Morning Herald*, Sydney; 23 October 2009, 'Land Free – in Return for a Good Feed', *Sydney Morning Herald*, Sydney.

35 Ibid; see also 16 January 2009, 'Africans Cry Colonialism Over Korean Land-Rental Deal', *Sydney Morning Herald*, Sydney.

36 Zhang, D et al, 2007, 'Global Climate Change, War, and Population Decline in Recent Human History' 104 *Proceedings of the National Academy of Sciences of the United States of America* 19214.

37 Ibid.

38 Ibid.

39 Ibid.

40 Ibid.

41 27 October 2005, 'Cow Politics', *New York Times*, New York.

42 FAO, 3–5 June 2008, High-Level Conference on World Food Security: The Challenges of Climate Change and Bioenergy, Rome.

43 Anderson, K and Martin, W, 2005, 'Agricultural Trade Reform and the Doha Development Agenda' 28 *The World Economy* 1301 at 1309.

44 Williams, B, 16 December 2009, Email to Ben Saul, on file with the authors.

45 See, eg, Group of 77, 6 April 2009, *Statement to the UN General Assembly* The Global Food Crisis and the Right to Food, New York; UN Human Rights Council, 10 September 2008, 'Human Rights Council Discussed Right to Food in the Context of the Global Food Crisis', Press Release.

46 FAO, 2009, *The State of Food Insecurity in the World*, FAO, Rome.

47 Intergovernmental Panel on Climate Change (IPCC), 2007, *Fourth Assessment Report: Climate Change 2007*, Cambridge University Press, New York.

48 United Nations Educational, Scientific and Cultural Organization (UNESCO), 2006, *Water: A Shared Responsibility*, World Water Development Report 2 (WWDR2), 4th World Water Forum, Mexico.

49 World Health Organization (WHO) and United Nations Children's Fund (UNICEF), 2007, *Meeting the MDG Drinking Water and Sanitation Target: The Urban and Rural Challenge of the Decade*, WHO–UNICEF, Geneva–New York.

50 Ibid.

51 UN Secretary-General Ban Ki Moon, 2007, *Water Security: Leadership and Commitment*, 1st Asia-Pacific Water Summit, Beppu.

52 Ibid.

53 Brown and Crawford, above n 25.

54 Ibid.

55 Ibid.

56 Selby, J, 2006, 'The Geopolitics of Water in the Middle East: Fantasies and Realities' 26 *Third World Quarterly* 329 at 343–44 in Podesta and Ogden, above n 9, p 123.

57 Brown and Crawford, above n 25.

58 Ibid.

59 Ibid.

60 Smith and Vivekananda, above n 2, p 14.

61 Ibid.

62 Brown and Crawford, above n 25.

63 See, eg, International Law Commission (ILC), 2008, 'Shared Natural Resources: The Law of Transboundary Acquifers - Draft Articles' UN Doc A/CN.4/L.724, Articles 4, 6, 10 and 12 respectively.

64 UN General Assembly, 14 December 1962, 'Permanent Sovereignty Over Natural Resources' UN GA Res 1803 (XVII).

65 Smith and Vivekananda, above n 2, p 3.

66 Ibid.

67 Dietz, above n 9, p 5.

68 August 2005, *Nature*; September 2005, *Science Magazine*.

69 WHO, 2005, *Fact Sheet: Climate and Health*, WHO, Geneva; see also WHO (Executive Board), 2008, *Climate Change and Health: Report by the Secretariat* EB122/4, WHO, Geneva; 2008, Res WHA51.29 and WHA61.19, 61st World Health Assembly, Geneva.

70 Ibid. Asia alone accounts for 80 per cent of the world's total deaths due to water-related disasters (2001–05); from 1960–2006 there were over 600,000 water-related disasters (floods, droughts, tsunamis, windstorms, landslides, storm surges, water-born diseases and epidemics).

71 Ibid.

72 Ibid.

73 Khasnis, AA and Nettleman, MD, 2005, 'Global Warming and Infectious Disease' 36 *Archives of Medical Research* 689; see also Senior, K, 2008, 'Climate Change and Infectious Disease: A Dangerous Liaison?' 8 *The Lancet Infectious Diseases* 92. For a more cautious assessment, see Zell, R, 2004, 'Global Climate Change and the Emergence/Re-emergence of Infectious Diseases' 293 *International Journal of Medical Microbiology* 16.

74 Lancet and University College London Institute for Global Health Commission, 2009, 'Managing the Health Effects of Climate Change' 373 *The Lancet* 1693 at 1694.

75 A term used by Michael Klare, 'Global Warming Battlefields: How Climate Change Threatens Security' 106 *Current History* 355 at 360.

76 See, eg, the popular discourse reflected in Dyer, above n 7. For a clear analysis of how this discourse has been perpetuated, see Lorraine Elliott, 'Climate Migration and Climate Migrants: What Threat, Whose Security?' in McAdam (ed), above n 7, p 175.

77 Bangladesh Institute of International and Strategic Studies/Saferworld, June 2010, 'Climate Change and Security in Bangladesh', p 19.

78 Ibid.

79 Ibid; Authors' interview with Rizwana Hasan (Bangladesh Environmental Lawyers Association (BELA)), Dhaka, 16 June 2010; World Bank Office, Dhaka, 2007, 'Improving Living Conditions for the Urban Poor', p 42; High Court of Bangladesh, Decision of 18 December 2008, Writ Petition No 9763/2008 filed by ASK; High Court of Bangladesh, Writ Petition No 1167/2008; ASK, 2009, 'Human Rights in Bangladesh 2008', pp 134–38; ASK, 2009, 'Annual Report 2008', p 12.

80 Authors' interview with Senior Official, Department of Environment, Ministry of Environment and Forests, Bangladesh, Dhaka, 14 June 2010; Authors' interview with Rizwana Hasan, BELA, ibid; World Bank Office Dhaka, ibid, p xvii, ch 5.

81 Authors' interview with Abul Kalam Azad (Research Director, Bangladesh Institute of International and Strategic Studies), Dhaka, 15 June 2010; Bangladesh Institute of International and Strategic Studies/Saferworld, above n 77, p 19.

82 Ibid.

83 Landlessness also has certain economic disadvantages, such as the inability to obtain credit (for instance, to start a business) if there is no land to secure the loan: see, eg, Abrar, CR and Azad, SN, 2004, *Coping with Displacement: Riverbank Erosion in North-West Bangladesh*, University Press Ltd Dhaka, pp 127–28; Seeley, J, et al, 2009, '"The Family is Suffering": Challenges Faced by Migrants' Families Who Stay Behind in a Village in Rural Northwest Bangladesh' in Chowdhury, RA and Seeley, J (eds), *Social Protection and Livelihoods: Marginalised Migrant Workers of India and Bangladesh*, University Press Ltd, Dhaka, p 49.

84 Shahadat Hossein, 2008, 'Rapid Urban Growth and Poverty in Dhaka City' 5 *Bangladesh e-Journal of Sociology* 1 at 19.

85 ASK, 'Human Rights', above n 79, pp 139–40.

86 UN Habitat, 'Case Study: Dhaka's Vulnerability to Climate Change, Backgrounder' in UN Habitat, 2008, 'State of the World's Cities 2008/2009'.

87 Ibid.

88 Ibid.

89 German Advisory Council on Global Change, 2007, 'World in Transition: Climate Change as a Security Risk', p 3; Campbell et al, above n 6, p 57; Podesta and Ogden, above n 9, p 118; Smith and Vivekananda, above n 2.

90 Authors' interview with Rizwana Hasan, BELA, above n 79; Authors' interview with Sultana Kamal, ASK, 'Human Rights', above n 79.

91 There is a rise in Islamic non-governmental organisations (NGOs) in some areas which promote religious values and run programs mainly for men. Abrar and Azad suggest that these well-funded NGOs could better assist communities by widening the focus of their programs: Abrar and Azad, above n 83, pp 104–05.

92 Authors' interview with Major General Golam Mohammad (Director General of Bangladesh Institute of International and Strategic Studies), Dhaka, 15 June 2010.

93 Hussain, MS, 'Terrorism in South Asia: Ramifications in the Internal and External Security of States: A Bangladesh Perspective' in Osmany, MR and Kabir MH (eds), 2007, *Global War on Terror: Bangladesh Perspective*, Academic Press and Publishers Library Dhaka, pp 130–31; see also Ollapally, DM, 2009, *The Politics of Extremism in South Asia*, Cambridge University Press, Cambridge, pp 182–83, 203.

94 Podesta and Ogden, above n 9, p 117; Campbell et al, above n 6, p 57; Barnett, above n 8, p 8; European Council, above n 4, p 7; Smith and Vivekananda, above n 2, p 16; Barnett, J, 2003, 'Security and Climate Change' 13 *Global Environmental Change* 7 at 8.

95 Authors' interview with Major General Golam Mohammad, above n 92.

96 Ollapally, above n 93, p 193.

97 Khan in, 23 April 2010, 'Target Practice' *New Age Extra* 11, available at <http://www.newagebed.com/2010/apr/23/xtra_cover.html>.

98 Kumar, A, 28 August 2006, 'Indo-Bangladesh Border Dispute Demands Urgent Attention', available at <http://www.southasiaanalysis.org/%5Cpapers20%5Cpaper1931.html>.

99 Authors' interview with Ahmed Swapan Mahmud (Executive Director, VOICE), Dhaka, 14 June 2010.

100 Ollapally, above n 93, p 198.

101 Shamshad, R, 2008, 'Politics and Origin of the India-Bangladesh Border Fence', pp 3–7; Ollapally, above n 93, pp 198–99; Rahman, R, 2007, 'Bangladesh's Fight against Terror: Lessons Learnt and Future Perspectives' in Osmany and Kabir (eds), above n 93, pp 13–16.

102 Behera, A D, 2008, *Violence, Terrorism and Human Security in South Asia*, University Press, Dhaka, p 63.

103 Hussain, above n 93, p 99.

104 Ibid, p 101.

105 Amin, ATM, 2009, 'Regional Security Environment and the Menace of Terrorism' in Ahmed, I (ed), *Terrorism in the 21st Century: Perspectives from Bangladesh*, University Press Ltd, Dhaka, pp 103, 107.

106 Behera, above n 102, pp 130–31.

107 Lama, MP, 2010, *Human Security in India: Discourse, Practices and Policy Implications*, ULP Dhaka, p 33.

108 According to the US National Snow and Ice Data Center in US National Intelligence Council, above n 21, p 53.

109 See, eg, Chatham House International Law Discussion Group, 2008, *The Arctic and Climate Change,* Meeting Summary, Chatham House, London.

110 Bowcott, O, 15 January 2009, 'Bush Urges US to Stake Claim to Arctic Territory in Last-Gasp Energy Grab', *The Guardian*, London; Associated Press, 4 October 2007, 'As Arctic Ice Melts, US Moves to OK Sea Treaty', *MSNBC*, New York.

111 See, eg, Rayfuse, R, 2007, 'Melting Moments: The Future of Polar Oceans Governance in a Warming World' 16 *Review of European Community and International Environmental Law* 196.

112 Paskal, above n 22, p 3.

113 UN Security Council debate, above n 3, p 3.

114 Barnett, above n 8, p 4.

115 ur Rashid, H, 17 May 2006, 'Is India taking advantage?', *Daily Star*, London in Paskal, above n 22, p 4.

116 European Council, above n 4.

117 Caron, D, 'Climate Change, Sea Level Rise and the Coming Uncertainty in Ocean Boundaries: A Proposal to Avoid Conflict' in Hong, S and Van Dyke, JM (eds), 2008, *Maritime Boundary Disputes, Settlement Processes, and the Law of the Sea*, Brill, Leiden, p 14.

118 European Council, above n 4.

119 See Elliott, above n 7.

120 German Advisory Council, above n 89, p 13; ADF, 2007, *Climate Change, The Environment, Resources and Conflict*, ADF, Canberra in Pearlman and Cubby, above n 6; Podesta and Ogden, above n 9, pp 132–33.

121 UN Security Council debate, above n 3; see also Sindico, F, 2007, 'Climate Change: A Security (Council) Issue?' 1 *Carbon and Climate Law Review* 29.

122 Draft Resolution, 27 October 2008, UN Doc A/63/L.8 sponsored by Armenia, Australia, Austria, Belgium, Bulgaria, Canada, Chile, Cyprus, Czech Republic, Denmark, Estonia, Fiji, Finland, France, Germany, Greece, Hungary, Iceland, Ireland, Israel, Italy, Kiribati, Latvia, Lithuania, Luxembourg, Maldives, Malta, Marshall Islands, Micronesia (Federated States of), Monaco, Nauru, Netherlands, New Zealand, Palau, Papua New Guinea, Philippines, Poland, Portugal, Romania, Samoa, Seychelles, Slovakia, Slovenia, Solomon Islands, Spain, Sweden, Tonga, Turkey, Tuvalu, United Kingdom of Great Britain and Northern Ireland and Vanuatu.

123 Asian Development Bank (ADB), *Water Financing Program 2006-2010 – Water for All: The Next Five Years*, ADB, Manila.

124 World Bank, 2006, *Clean Energy and Development: Towards an Investment Framework* DC2006-0002, World Bank, Washington DC, annex K.

125 Mainhardt-Gibbs, H, 2009, *World Bank Energy Sector Lending: Encouraging the World's Addiction to Fossil Fuels*, Information Centre Brief, World Bank, Washington DC.

126 Miles, K, 2008, *International Investment Law and Climate Change: Issues in the Transition to a Low Carbon World*, Paper presented at New Horizons of International Economic Law, The Inaugural Conference of the Society of International Economic Law, Geneva; see also Miles, K, 'Sustainable Development, National Treatment and Like Circumstances in Investment Law' in Cordonier Segger, M, Newcombe, A and Gehring, M (eds), 2011, *Sustainable Development in World Investment Law*, Kluwer, The Hague, p 261.

127 Vidal J, 21 April 2009, 'Climate Change Will Overload Humanitarian System, Warns Oxfam' *The Guardian*, London.

Glossary

Agribusiness: large-scale commercial agriculture.

Anthropogenic: caused by human activity.

Australian Academy of Science: comprised of more than 400 distinguished Australian scientists, elected to the Academy by the existing fellows.

Australian Greenhouse Office: the Australian Government office responsible for reducing greenhouse gas emissions, including by administering energy-efficiency programs.

Australian Human Rights Commission: the Australian Government agency responsible for promoting the protection of human rights.

Baseline and credit: a type of emissions trading scheme that does not limit total emissions, but which assigns participants tradable credits (or offsets) for reducing emissions below a baseline level.

Biochar: charcoal made of biomass, the manufacture of which is a form of carbon capture and storage.

Biofuel: a fuel derived from plant matter.

Boat people: a colloquial term referring to asylum seekers who come to Australia by boat without a visa.

Business as usual warming: global warming that will occur if increases in greenhouse gas emissions and economic growth continue at current rates.

Calthrate: a compound in which molecules of one component are fixed within the crystal structure of another.

Cap and trade: a type of emissions trading scheme that limits the total allowed emissions.

Carbon capture: technology that captures greenhouse gas emissions from burning fuel, including coal or oil, before they enter the atmosphere.

Carbon credit: a permit that allows a country or a company to produce a fixed amount of greenhouse gas, which can be traded in an emissions trading scheme if it is not used.

Carbon footprint: the total greenhouse gas emissions attributable to an activity or product.

Carbon intensity: the relative average greenhouse gas emissions from a particular activity.

Carbon leakage: a phenomenon by which efforts to reduce greenhouse gas emissions in one country cause an increase in emissions in another country.

Carbon Pollution Reduction Scheme (CPRS): a proposed Australian greenhouse gas emissions trading scheme that failed to gain sufficient support in the Australian Parliament.

Carbon tax: a tax on goods and services proportional to the greenhouse gas emissions that they generate.

Causation: various areas of law have developed technical rules for determining when a person or entity can be held legally responsible for causing harm to others.

Certified Emission Reductions: carbon credits issued for emission reductions achieved under the Kyoto Protocol's Clean Development Mechanism.

Chlorofluorocarbons (CFCs): a class of compounds, many of which have been widely used as refrigerants, propellants and solvents, that are both ozone depleting and greenhouse gases.

Clean coal: the use of technology such as carbon capture and sequestration to reduce the amount of greenhouse gas that enters the atmosphere from burning coal to generate electricity.

Climate change: altered global and regional climate patterns caused by global warming, including for example: changed rainfall patterns, rising sea-levels, desertification, extreme weather events and ocean acidification.

Climate denial: denial of the scientific consensus surrounding climate change, including the view that climate change science is a hoax or conspiracy.

Climate-displaced people: people forced to move from their region of origin by climate change.

Climate wars: theoretical armed conflicts caused by climate change.

Commitment period: a period of time specified in an agreement between countries, during which countries undertake to perform certain measurable activities, such as reducing carbon emissions by specified amounts.

Comparable effort: an approach to allocating emission allowances between countries based on the relative difficulty of reducing emissions, contrasted with the contraction and convergence approach.

Complementary protection: visas or other protections or benefits afforded to people who have been displaced from their country of origin, or face a risk of harm if they return, but who do not meet the legal definition of a refugee. Also known as subsidiary protection.

Contiguous zone: a maritime zone adjacent to a state's territorial sea, up to 44 km from the shoreline, over which a state may exercise control for the purpose of enforcing its laws.

Contraction and convergence: an approach to allocating emission allowances between countries by setting a global target and reducing national emissions over time until all countries' per capita emissions are equal. This is contrasted with the comparable effort approach.

Direct action: measures designed to reduce global warming without using a market-based approach such as a carbon tax or emissions trading scheme, for example by promoting energy efficiency or use of renewable energy sources.

Dirty fuel: a fuel that is particularly carbon intensive.

El Nino: a complex pattern of atmospheric and ocean currents that occurs intermittently, between contrasting La Nina patterns, in the Pacific region.

Electric arc furnace: a furnace that has a smaller carbon footprint than a traditional blast furnace because it uses less energy and can make steel from scrap metal.

Emission Reduction Units: the trading unit for carbon credits under the Kyoto Protocol, representing a reduction of one tonne of CO_2 or equivalent greenhouse gas emissions.

Emissions trading scheme: a regulatory scheme that assigns carbon credits to participants and allows them to trade them, producing market incentives to reduce emissions.

Energy efficiency target: a target for the amount of energy a particular technology, such as a car or appliance, should consume while doing a particular task.

Exclusive Economic Zone: the zone recognised under international maritime law, up to 370 km from the shoreline, in which a state has special rights to exploit marine resources.

Extreme weather event: unusually severe weather, such as extreme heat or cold, severe hydro-meteorological events, flooding or bush fires.

Forced displacement: coerced movement of people away from the region in which they usually live.

The Garnaut Review: *The Garnaut Climate Change Review*, a 2008 report by Professor Ross Garnaut, commissioned by the Australian Commonwealth and State and Territory governments.

General circulation model: a mathematical model of the circulation of the Earth's atmosphere or ocean, which is the basis for computer programs that simulate the atmosphere, used for weather forecasting and projecting climate change.

Geo-engineering: technology used to remove carbon dioxide from the atmosphere or otherwise artificially cool the planet.

Global warming: long-term gradual increase in the average temperature of Earth's atmosphere and oceans. Although periods of global warming and cooling have occurred throughout the Earth's history the term often refers to anthropogenic global warming – warming caused by increased concentrations of greenhouse gases in the atmosphere due to human activity.

Greenhouse gases: gases that absorb and emit heat energy, including carbon dioxide, methane, and nitrous oxide. Atmospheric greenhouse gases cause warming known as the greenhouse effect.

Human security: a concept of security focused on human well-being rather than the security of the state.

Hybrid climate-biosphere models: a mathematical model of the atmosphere and other aspects of the biosphere such as plant growth and water and carbon cycles. These are the basis of computer programs that simulate the interaction between these factors (such as increased plant growth if atmospheric carbon dioxide increases) in projecting climate change.

Hydrochlorofluorocarbons (HCFCs): a sub-class of CFCs that are less-ozone depleting than other CFCs.

Hydro-meteorological disasters: natural disasters involving rain or snow, such as flooding, hurricanes, typhoons, cyclones, blizzards or mud-slides.

Intergovernmental Panel on Climate Change: An intergovernmental organisation established by the United Nations Environment Program and the World Meteorological Society, with 194 member governments. Thousands of scientists contribute on a voluntary basis to the IPCC's work reviewing scientific information about climate change.

Internally displaced persons: persons forced to move away from the region in which they usually live, but who remain within their country of origin.

Market-based approach: an approach to reducing greenhouse gas emissions that imposes a price on carbon and relies on economic mechanisms to reduce emissions, including a carbon tax or emissions trading scheme.

Market failure: a situation in which individuals' pursuit of self-interest in a market leads to results that are not optimal for the market participants as a whole, and may not be optimal for individuals.

Mitigation: Actions taken to decrease global warming by reducing the concentration of greenhouse gases in the atmosphere.

Most favoured nation principle: a principle in international trade law. The members of the World Trade Organization accord each other most favoured nation status, granting each other member equally the most-advantageous application of their domestic trade law such as tariffs (import or export taxes) and import quotas.

National treatment principle: a principle in international trade law mandating equal treatment of imported and domestically produced goods.

New South Wales Greenhouse Gas Abatement Scheme: An emissions trading scheme under which electricity retailers pay financial penalties if they do not meet greenhouse gas abatement targets.

Non-refoulement: a principle of international law under which a refugee must not be returned to a country in which they are likely to be persecuted.

Ocean acidification: decrease in the pH of the world's oceans caused by uptake of carbon dioxide from the Earth's atmosphere.

Ocean fertilisation: introduction of nutrients to the upper ocean to increase the growth of phytoplankton (microscopic ocean plants) and thereby increase uptake of carbon dioxide from the atmosphere.

Paleoclimate record: the evidence of the climate during Earth's history found in nature, including in rock, sediment, ice sheets, tree rings, coral, shells and microfossils.

Precautionary principle: a principle of environmental law and policy that states that measures should be taken to avoid harm to the environment where there is evidence, but not necessarily full scientific consensus, regarding the harm.

Protectionism: protection of domestic industries from foreign competition, for example by imposition of tariffs or import quotas.

Protocol: a protocol is another name for a treaty under international law.

Ratification: the process by which a state formally enters into binding international legal obligations under a treaty.

Renewable energy: energy from sources that are naturally replenished, including solar, wind, tidal, geothermal and biofuels.

Slow onset disasters: natural disasters that emerge gradually, such as drought or desertification.

Socio-economic resilience: an individual's or community's ability to access and use social or economic resources to adapt to challenges.

Spectroscopic theories: scientific theories about the interaction of radiation energy (including heat) and matter (including gas molecules), which explain how greenhouse gases cause global warming, and how much warming they will cause.

Smart grid: an electricity grid that uses digital communication to reduce the total amount of electricity consumed, for example by communicating to consumers about electricity use and pricing, or ensuring that appliances and machinery run during times of low electricity demand.

State: the formal legal name for a country under international law.

Stateless: a person who has no legal nationality.

The Stern Review: *The Stern Review on the Economics of Climate Change*, a 2006 report by Sir Nicholas Stern, commissioned by the British government.

Subsidiary protection: visas or other protections or benefits afforded to people who have been displaced from their country of origin, or face a risk of harm if they return, but who do not meet the legal definition of refugee. Also known as complementary protection.

Survival migration: an individual's or group's movement away from their region of origin in response to pressures that threaten their well-being and cannot be remedied without moving, such as conflict or climate change.

Tar sands: a type of unconventional oil deposit, in which dense oils are found mixed with sand, clay and water.

Technologism: the view that the response to climate change should rely primarily on using technology, rather than market mechanisms or behavioural change, to achieve mitigation and adaptation.

Territorial sea: coastal waters, up to 22 km from the shoreline, that are treated under international law as part of the sovereign territory of the state.

Trade liberalisation: the reduction of barriers to international trade including subsidies, tariffs (import or export taxes) and import quotas.

Tragedy of the commons: the idea, first defined by Garrett Hardin, that multiple individuals acting in self-interest with access to a common resource will deplete it, even if depletion is not in their collective long-term interest.

Transaction cost: the cost incurred in making an economic exchange. Participants in emission trading schemes will incur transaction costs, which affect the efficiency of the scheme.

Transnational crimes: organised crime coordinated across national borders.

Transnational security threats: diverse phenomena that may give rise to conflict, mass violence or other threats to national or human security, such as epidemics of infectious disease, international organised crime, terrorism, extreme weather events and climate change.

Treaty: a written agreement between two or more states that gives rise to binding obligations in international law. It may also be called a convention, agreement, covenant or protocol, among other terms used.

Unconventional oil: oil extracted using methods other than an oil well. Extracting unconventional oil is generally more environmentally destructive and energy intensive than extracting conventional oil, making unconventional oil's carbon footprint greater.

Urban design: the design of towns and cities, including considerations of infrastructure, housing development, architecture and landscape design.

Vehicle-to-grid (V2G): a smart-grid technology in which electric vehicles can sell the energy stored in their batteries back to the grid in periods of high demand.

Voluntarism: the view that countries should reduce greenhouse gas emissions voluntarily, rather than pursuant to international legal obligations.

Index